NUCLEAR WASTE
MANAGEMENT
ABSTRACTS

IFI DATA BASE LIBRARY

NUCLEAR WASTE MANAGEMENT ABSTRACTS

Richard A. Heckman and Camille Minichino

Lawrence Livermore National Laboratory
University of California
Livermore, California

IFI / PLENUM • NEW YORK-WASHINGTON-LONDON

Library of Congress Cataloging in Publication Data

Heckman, Richard A.
　Nuclear waste management abstracts.

　(IFI data base library)
　Bibliography: p.
　Includes indexes.
　1. Radioactive waste disposal—Abstracts., I. Minichino, Camille. II. Title. III. Series.
TD898.H43　　　　　363.7'28　　　　　81-19868
ISBN 978-1-4684-6137-4　　　ISBN 978-1-4684-6135-0 (eBook)
DOI 10.1007/978-1-4684-6135-0
　　　　　　　　　　　　　　　　　　　　　AACR2

© 1982 IFI/Plenum Data Company
Softcover reprint of the hardcover 1st edition 1982

A Division of Plenum Publishing Corporation
233 Spring Street, New York, N.Y. 10013

PREFACE

As we enter mid-1981, the Reagan administration is completing a review of U. S. nuclear waste management policy. Major revisions in the recently announced Carter administration policies are expected. Reagan is a strong supporter of civilian nuclear power and will probably encourage spent fuel reprocessing by the private sector. Meanwhile, the deep geologic disposal of defense nuclear waste in New Mexico moves ahead.

In the coming months, discussion and debate of U. S. radioactive waste management policies will intensify in the Congress, in the technical community, and among environmentalists and the public at large. An important element of the debate should be the scientific and technical issues of the safe disposal of radioactive wastes from both the civilian nuclear power fuel cycle and the defense fuel cycle, including naval pro-propulsion programs and nuclear weapons production.

The literature of waste management is voluminous, covering all aspects of the world-wide problem of safe disposal. The authors of this book have attempted to critically review this literature, selecting the more important reports to abstract. Our selection criteria were heavily influenced by considerations of policy issues and by our experiences in both the technical community and the regulatory environment. Our intent is to identify those reports we feel will contribute the most to the development of a national consensus on the safe disposal of existing and future nuclear wastes as yet another U. S. waste policy emerges in Washington.

The authors acknowledge the valuable assistance of Ms. Amelia Wilcox in preparing the manuscript.

CONTENTS

1. THE PROBLEM OF NUCLEAR WASTE MANAGEMENT

For almost 40 years, radioactive wastes have been generated by national defense programs, by the commercial nuclear power industry, and by various medical and research activities. Still, efforts to manage radioactive wastes have not been adequate, and no final, comprehensive plan exists for their disposal.

To understand the scope of the problem, we should consider the sources of waste from the nuclear fuel cycle; the temporal nature of the hazards from nuclear waste disposal relative to other natural hazards; current and projected waste inventories; and the regulatory framework for policies and decisions about nuclear waste disposal.

Figure 1 depicts schematically the total light water reactor (LWR) fuel cycle. Since the adoption in October, 1977, of a policy of nonproliferation of nuclear weapons, only the front end and the interim storage step of the back end of the fuel cycle are used. If future policies allow the fuel cycle to be closed, then the uranium and plutonium streams will be recycled as shown and high-level nuclear waste will be disposed in deep geological repositories. Under current policies, spent fuel is treated as a high-level waste and will be managed as such.

The waste streams generated from this cycle are shown in Figure 2. Both current policy and possible future reprocessing options are included. The fuel cycle steps are shown in the top row. This representation of the waste streams is useful because it clearly shows the commonality of the subsequent operations, irrespective of the waste generation step. The second row defines the type of waste generated. Some form of processing is required next, e.g., solidification, treatment and/or packaging, as shown in the third row. Some types of wastes will require interim storage at the processing facility, as shown in the fourth row. In the fifth row, the transportation step required to move the waste to the disposal or burial site is shown. Some types of wastes may require interim storage at the disposal or burial site as shown in the sixth row. Finally, the low-level wastes are buried in near surface sites and high-level/transuranic wastes will be disposed in deep geological repositories, as shown in the last row.

Figure 3 is a schematic representation of how very small amounts of radioactivity will eventually migrate to the biosphere from a deep geologic repository. Proper siting will allow the choice of repository geology most favorable to minimize the migration rate. The repository proper is mined out of a very "tight" layer between two very impermeable layers, or aquicludes. Since the repository site will probably be located in hyrologic basins, water-bearing formations, or aquifers, are shown above and below the aquicludes. Rainfall in mountains recharges the aquifers, producing the required hydraulic head, and water flows through the aquifers and eventually discharges into a surface water system, e.g., a stream or river, in the middle of the basin. These hydraulic gradients cause water to migrate through the repository layer. Eventually water penetrates into the waste canisters and dissolves or leaches radioactive material. This radioactivity begins a slow, tortuous migration toward the surface water system.

Current calculations suggest migration times of tens of thousands of years to hundreds of thousands of years, depending, e.g., on waste dissolution properties, hydrologic factors, and aquifer ion exchange properties, to the surface water systems in the middle of the basin. The potential radiological impacts to man are determined in large part by the human water usage patterns by the population living in equilibrium with surface water systems downstream from a repository. This is shown schematically in Figure 4.

A relative toxicity index (Figure 5) is a useful performance measure to understand the temporal nature of the hazard from nuclear waste and for a comparison with other natural hazards, e.g., from mercury ore, lead ore, and the original ore used to obtain the LWR fuel initially. It is generally accepted that after approximately 500 years, the fission products have decayed leaving only the transuranic elements and their daughters in the nuclear waste. Note that the index of the high-level waste is equal to that of an average uranium ore in approximately 20,000 years.

Figures 6 and 7 summarize health effects and types of radiation exposure to radioactive materials. These summaries are useful in analyzing radiological risks and the impacts of managing and disposing of nuclear wastes from the LWR fuel cycle.

The present inventory and generation rates of radioactive waste are given in Table 1. Note that although the volumes of the total defense and commercial high-level wastes are very different, they both contain approximately the same quantity of longlived fission products, or in other words, the same quantity of radioactivity. The quantity of spent fuel or high-level nuclear waste generated in the year 2000 can be estimated by a simple scaling of the generation rates and by noting that at present, the nuclear industry has an installed capacity of approximately 50 GWe.

Figure 8 outlines the regulatory framework within which discussion, debate, and decisions about waste management policy take place. A review of the history of waste management in the United States reveals a complex decision-making pattern (shown schematically in Figure 9) beginning several decades ago. Radioactive wastes produced through the Manhattan Project are still in temporary liquid storage. The National Research Council - National Academy of Science (NRC-NAS) working group issued its famous 1954 report which led to the development of a waste management policy of terminating ocean disposal and focusing research on deep underground disposal. The United States waste management program has been characterized by technical reviews leading to temporary policy changes and by experimental studies with abrupt terminations. Milestones in this history are shown in Figure 9.

In a message to Congress on February 12, 1980, then President Carter presented his proposal for establishing a comprehensive program for the management of all types of radioactive wastes, and recommended that a full-scale repository for both defense and commercial wastes be operational by the mid-1980's. The Reagan Administration is reviewing the issues again and is expected to announce a new waste management policy for the United States in 1981.

The problem of nuclear waste management is multifaceted. This book contains abstracts of literature covering all aspects of the problem: waste form and inventories; the transportation of waste; the technical issues involved in the selection of a suitable disposal site; waste repository construction and design; and the societal, political, and economic issues usually involved in decisions which command public attention and concern.

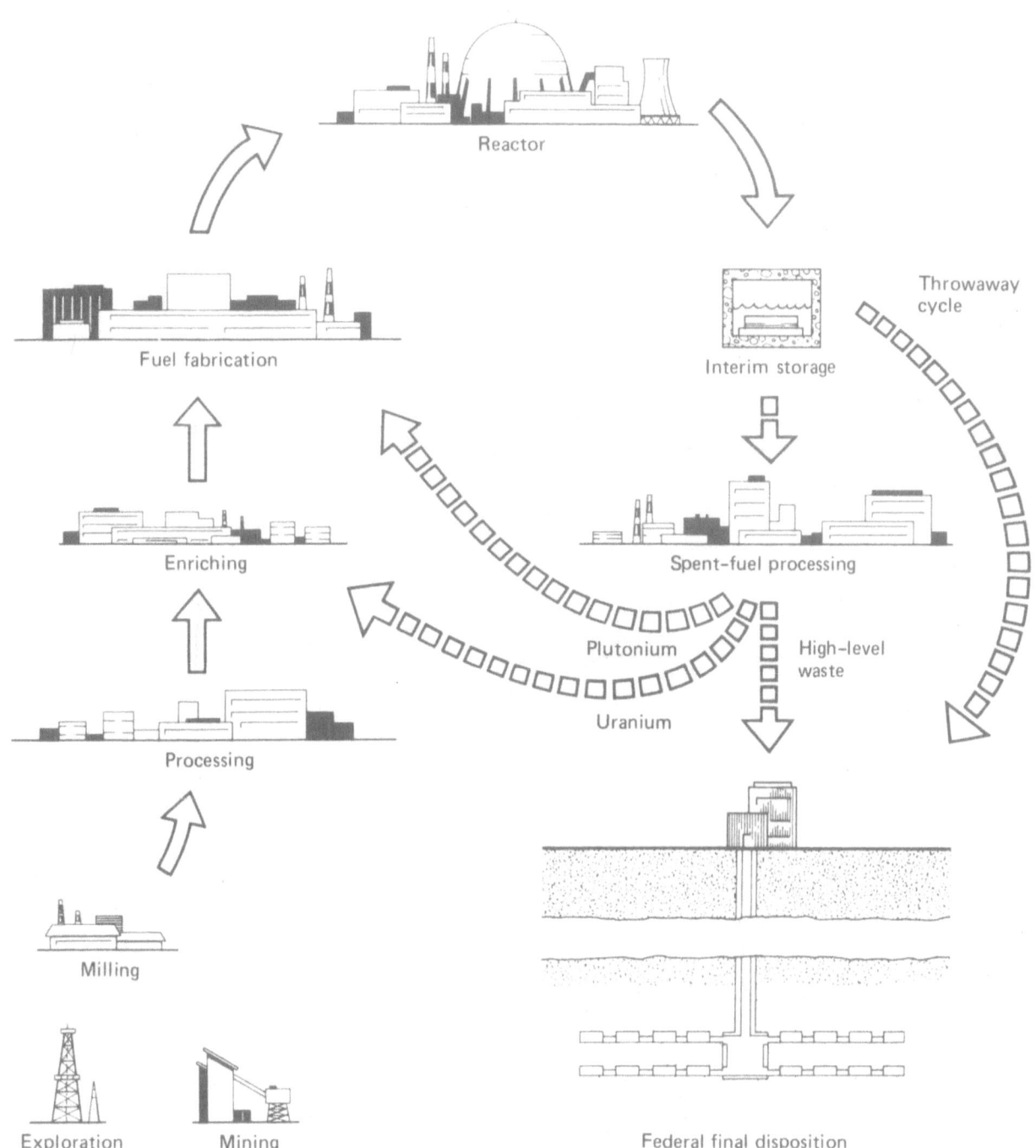

FIG. 1. Schematic representation of the LWR fuel cycle. Solid arrows show the fuel cycle
as it now exists; dashed arrows show the prospective "closed" fuel cycle that
depends on regulations to be issued by NRC. At present, spent fuel is kept in
interim storage, but future fuel cycles will use interim storage to permit spent
fuel to cool off before reprocessing. Future fuel cycles may require permanent
isolation of high-level wastes from either uranium or plutonium fuel cycles or
from spent fuel itself.

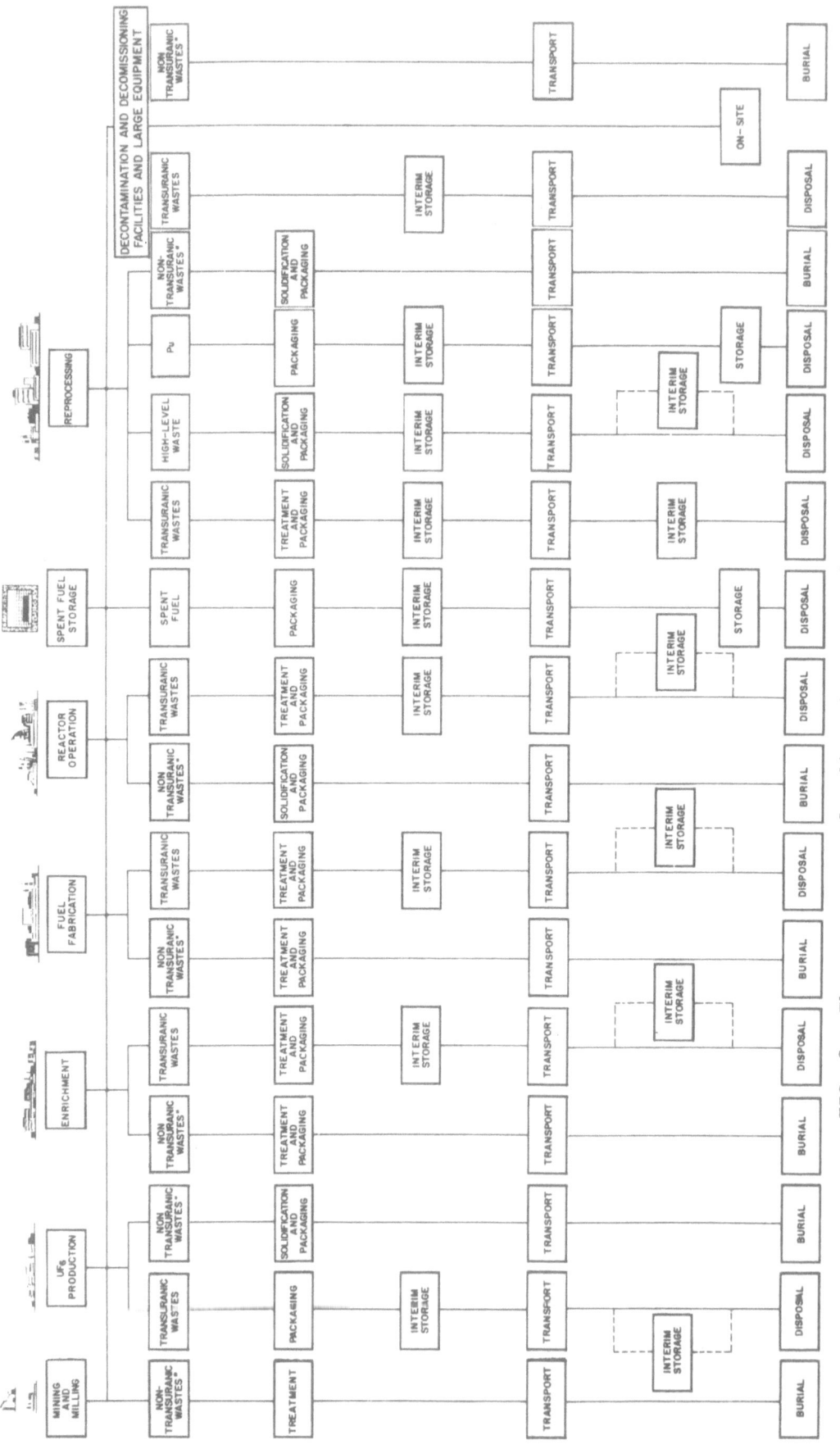

FIG. 2. The management of radioactive wastes from the LWR fuel cycle.

FIG. 3. Schematic representation of a deep geologic nuclear waste repository. Arrows show pathways by which small amounts of radioactivity will eventually migrate to the biosphere.

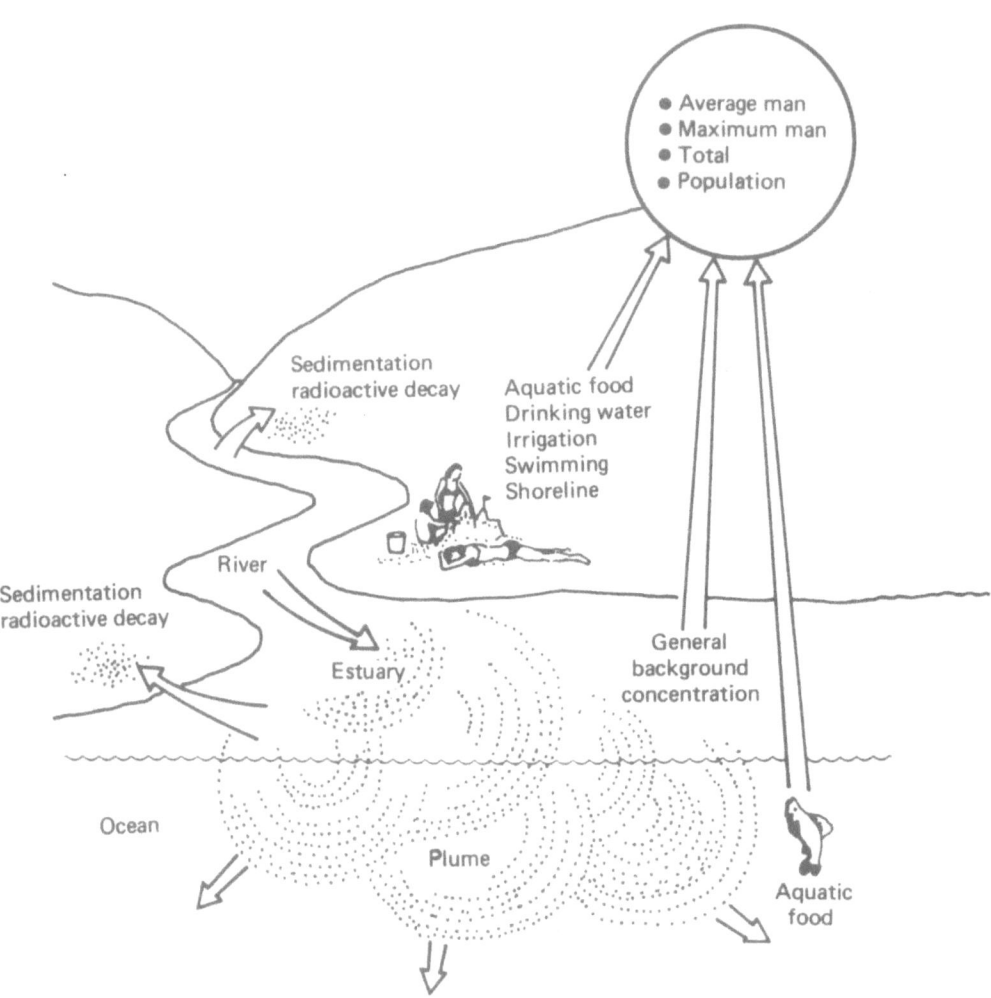

FIG. 4. Human water usage model. Potential radiological impacts to humans are
greatly influenced by water usage patterns

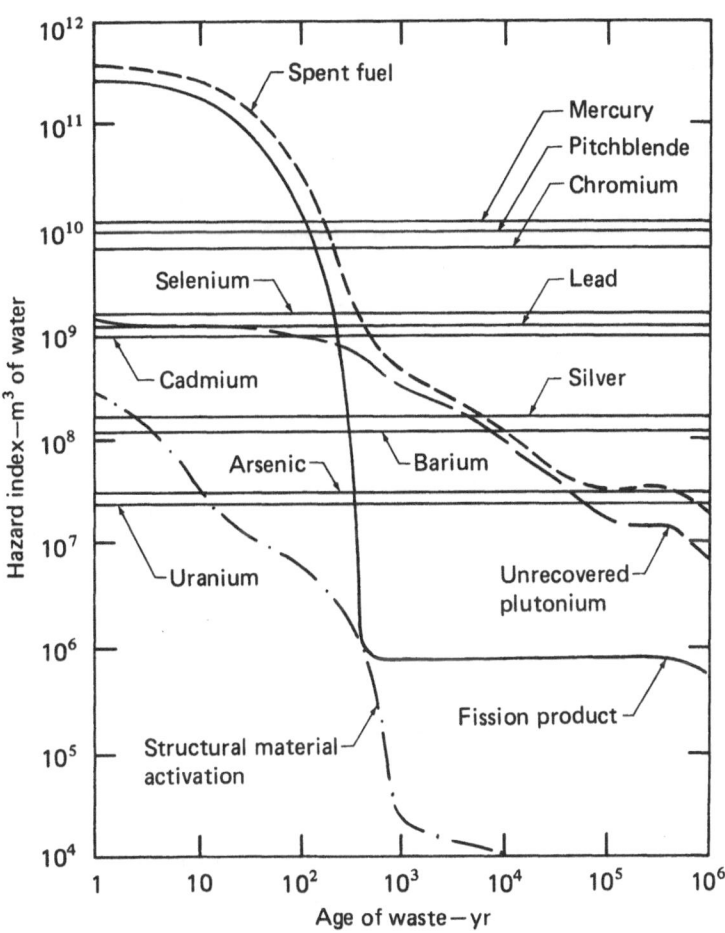

FIG. 5. This chart shows how the hazards associated with the various components of spent fuel from one year's operation of a 1000-MW e pressurized water reactor (one reference reactor year, or 1 RRY) decay with time and how these hazards compare with those from an equal volume of average ores of common toxic elements. The hazard index is the volume of water in which a substance would have to be dispersed to render it harmless.

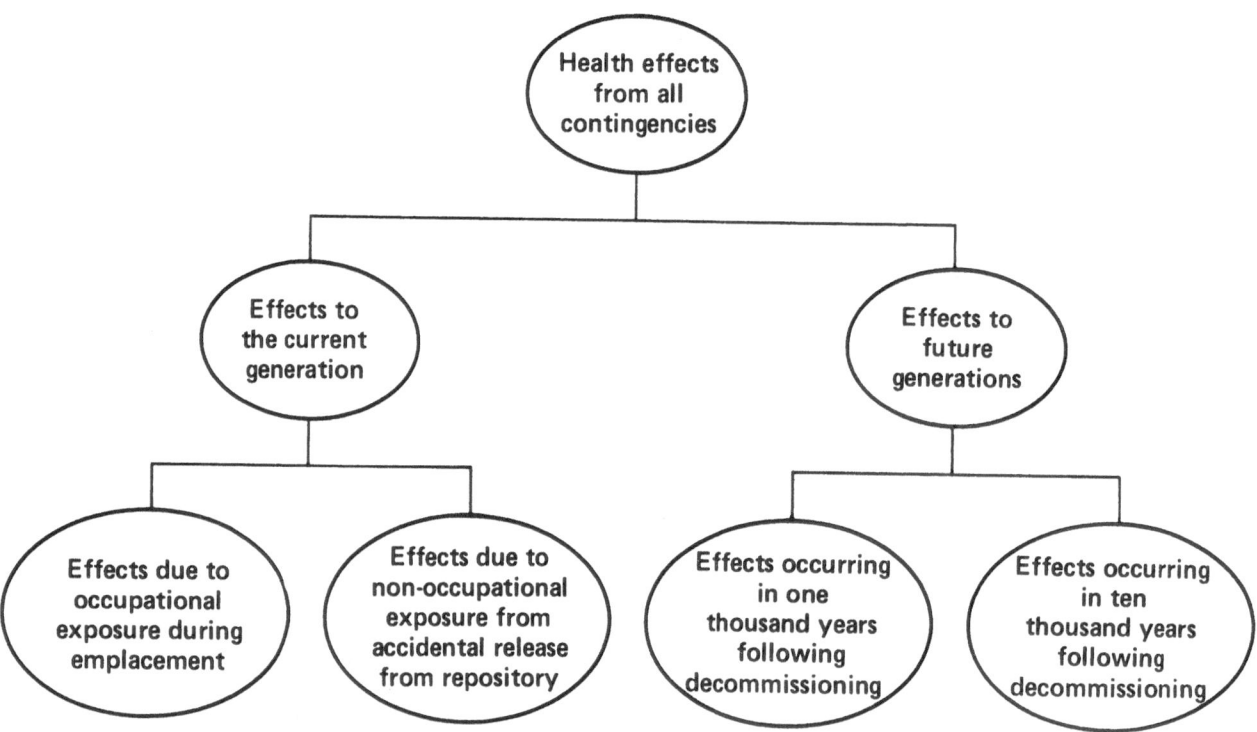

FIG. 6. Health effects from nuclear wastes.

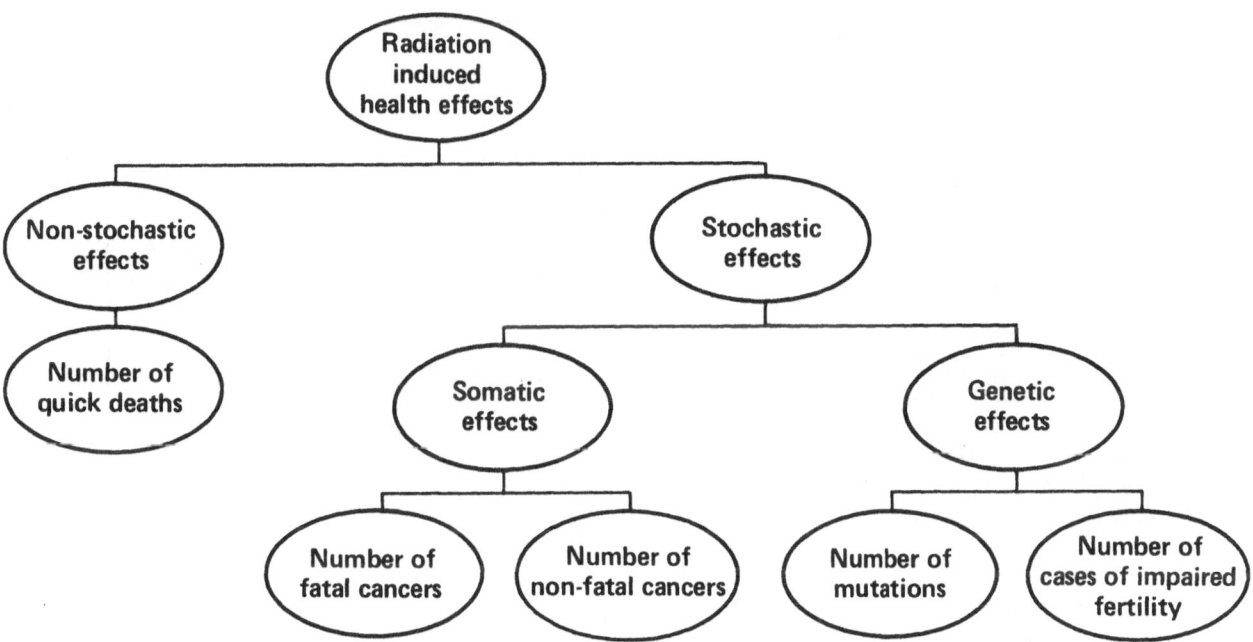

FIG. 7. Types of radiation exposure.

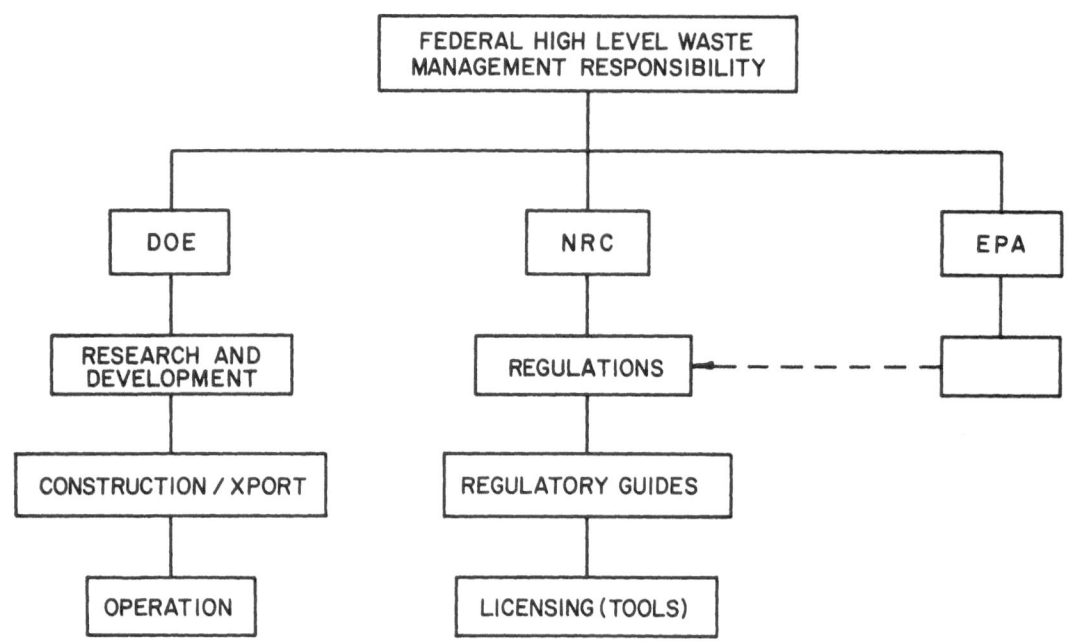

FIG. 8. The waste management responsibility is shared by government agencies.

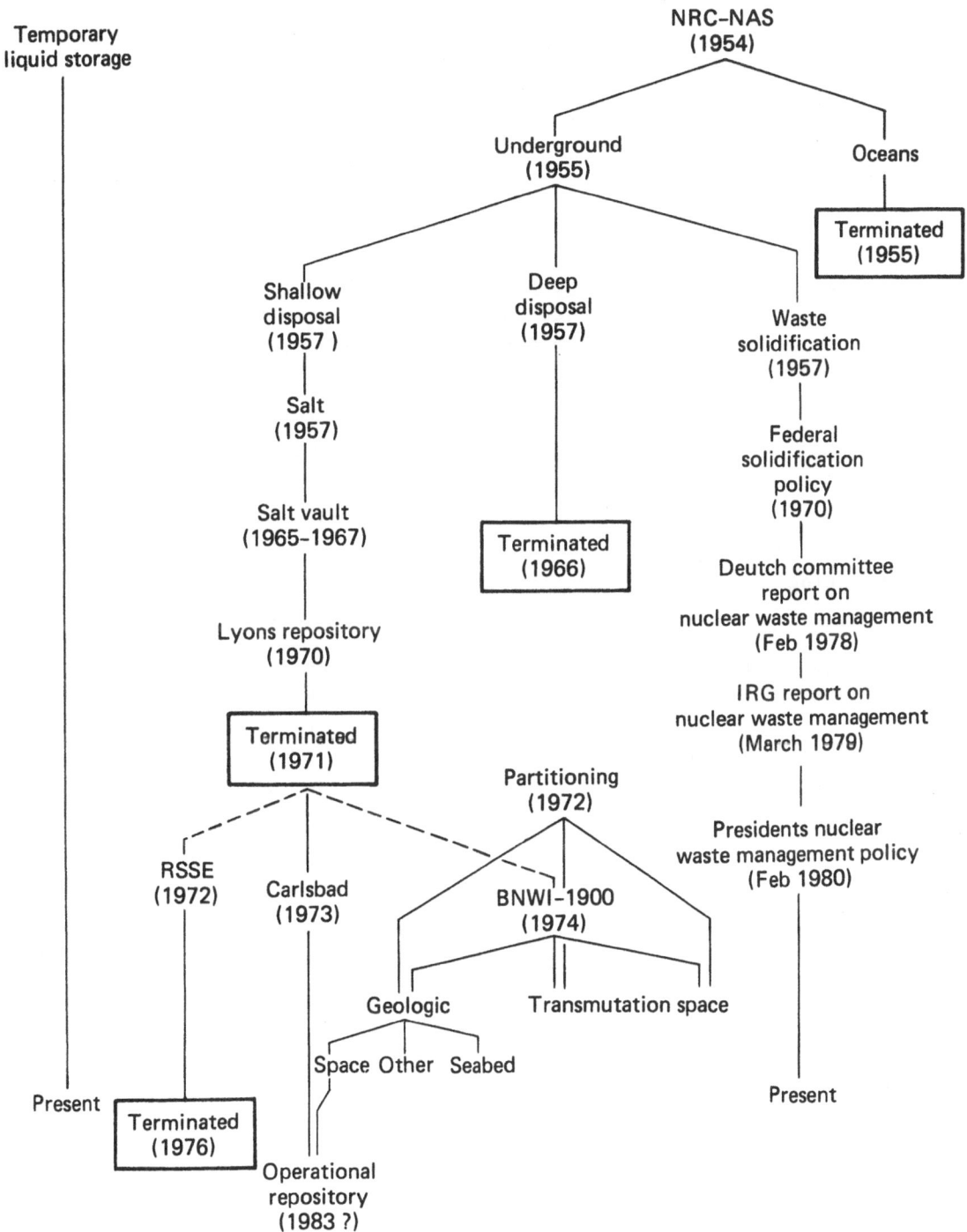

FIG. 9.   The nuclear waste management program of the United States has a complex
history of technical reviews, policy changes, and experimental studies.

TABLE 1. Present inventory and generation rate of radioactive
         wastes from either spent fuel or its equivalent in
         high-level wastes after reprocessing

Sources	Amount, m^3	
	Spent fuel	High-level waste
Existing		
Military	280,000 (mixed)	
Commercial	~3500	2300
	(both exist now)	
Generation rates		
Present industry (1 year)	2200 or	512
One reactor for life (30 years)	1050 or	240

2. GUIDE TO THE ABSTRACTS

Chapter 3 of this book contains abstracts of books, reports, and papers important in the history of nuclear waste management. It is organized according to four aspects of the problem:

  A. Waste Form and Inventories

  B. Transportation of Waste

  C. Site Selection and Repository Design

  D. Societal, Political, and Economic Issues of Waste Management

Section E is a list of bibliographies; Section F is a list of journals, progress reports, and other occasional publications. Only English language publications are included. This book also includes a glossary of terms relevant to nuclear waste management, a key word index, a title index, and an author index.

When more than one category seems appropriate for an abstract, cross-referencing is provided. Abstracts of publications especially significant in the history of nuclear waste management are marked with an asterisk.

Acronyms used in this book are as follows:

AEC - U. S. Atomic Energy Commission

ANS - American Nuclear Society, Inc., Hinsdale, IL

ASCE - American Society of Civil Engineers

BPNL - Battelle Pacific Northewest Laboratory, Richland, WA

CEC - Commission of the European Communities

DOE - U. S. Department of Energy, Washington, D. C.

EA&T - Engineering Analysis and Test Co., Inc., Marina del Rey, CA

EPA - U. S. Environmental Protection Agency

ERDA – U. S. Energy Research and Development Administration, Washington D. C.

GAO – General Accounting Office

IAEA – International Atomic Energy Agency, Vienna, Austria

IECO – International Engineering Company, San Francisco, CA

IRG – Interagency Review Group

KBS – Nuclear Fuel Safety Project, Sweden

LANL – Los Alamos National Laboratory, Los Alamos, NM

LBL – Lawrence Berkeley Laboratory, Berkeley, CA

LLNL – Lawrence Livermore National Laboratory, Livermore, CA

LWR – Light-Water Reactor

MIT – Massachusetts Institute of Technology, Cambridge, MA

NEA – Nuclear Energy Agency, Organization for Economic, Cooperation and
      Development, Paris, France

NRC – U. S. Nuclear Regulatory Commission

NRPB – National Radiological Protection Board, Harwell, Didcot, Oxfordshire, England

NTIS – National Technical Information Service, Springfield, VA

NTS – Nevada Test Site

NWTS – National Waste Terminal Storage (Program)

OECD – Organization for Economic Cooperation and Development, Nuclear Energy
       Agency, Paris, France

ONWI – Office of Nuclear Waste Isolation, Columbus, OH

ORNL – Oak Ridge National Laboratory, Oak Ridge, TN

OWI – Office of Waste Isolation, Oak Ridge, TN

SAI – Science Applications, Inc., Palo Alto, CA

SKBF – Swedish Nuclear Fuel Supply Co., Sweden

SLA – Sandia Laboratories, Albuquerque, NM

SRL – Savannah River Laboratory, Aiken, SC

TERA – Teknekron Energy Resource Analysts, Berkeley, CA

TIC – Technical Information Center, Oak Ridge, TN

UCB – University of California at Berkeley, Berkeley, CA

WIPP – Waste Isolation Pilot Plant, Eddy County, NM

3. ABSTRACTS

A. Waste Form and Inventories

B. Transportation of Waste

C. Site Selection and Repository Design

D. Societal, Political, and Economic
    Issues of Waste Management

E. Bibliographies

F. Journals and Progress Reports

A.   WASTE FORM AND INVENTORIES

A.1   Final Purification and Concentration of Americium/Curium Separated From High-Level Reprocessing Waste

H. Haug, Kernforschungszentrum, Karlsruhe, Germany

Journal of Radioanalytical Chemistry 21, April 1973, p. 187

This article discusses a cation-exchange cycle developed for the recovery and concentration of the Am/Cm product from a DTPA/lactic aid solution used in an extraction process for the isolation and separation of the actinides from lanthanide fission products.

KEY WORDS: actinides, reprocessing

A.2   Management of Plutonium – Contaminated Solid Wastes

NEA

Proceedings of the NEA Seminar – Marcoule, France 1974

The seminar was organized to review the management of plutonium-contaminated solid wastes based on experience acquired by installations which handle large quantities of plutonium. The proceedings represent a synthesis of current practices and research and development in this area.

KEY WORDS: solid wastes, plutonium

A.3   Incentives for Partitioning High-Level Waste

H. Burkholder, M. Cloninger, D. Baker, and G. Jansen, BPNL

BNWL-1927, November 1975

This report presents incentives for separating and eliminating various elements particularly transuranics from radioactive waste prior to final geologic storage. Assumptions used calculate doses, especially those concerning the transport of radioactivity from the geologic storage site to man, are detailed. The methodology in the study can be extended to evaluate any combination of waste form and site as waste management options.

KEY WORDS: environment, transport, dose

A.4   Bituminization of Radioactive Wastes at the Nuclear Research Center Karlsruhe – Experience from Plant Operation and Development Work

W. Hild, W. Kluger, and H. Krause, Gesellschaft fur Kernforschung mbh, Karlsruhe, Germany

KFK-2328, May 1976

This report summarizes experience and results gained in radioactive waste bituminization in radioactive waste bituminization

17

at the Nuclear Research Center, Karlsruhe, both in plant operation and research and development activities. Details of the process are given.

KEY WORDS: Germany

A.5   Metamict Mineral Alteration: An Implication for Radioactive Waste Disposal

R. Ewing, University of New Mexico, Albuquerque, NM

Science, 192, June 25, 1976, p. 1336

A possible method for evaluating the long-term stability of silicate or borate glasses by examining metamict minerals for alteration effects. Preliminary data is presented to suggest the inappropriateness of glass as a medium for radioactive waste disposal due to its susceptibility to alteration.

A.6   Proceedings of the Symposium on Waste Management, Tucson, Arizona, October 1976

This is a collection of papers on various aspects of waste management, with the theme of placing the hazards of radioactive waste in perspective with those wastes resulting from the generation of electrical energy by other sources. Sessions were focussed on waste sources and characteristics, geological disposal, and international programs and policies.

KEY WORDS: waste management programs

A.7   Durability of Containers for Storing Solidified Radiactive Wastes

C. Angerman and W. Ranken, SRL

DP-MS-76-66, 1976

This report discusses the durability of five alloys considered for use as solidified radioactive waste canisters. The evaluation includes parameters such as cost, oxidation resistance, strength, and compatibilities under conditions expected during 100-year storage.

A.8   Management of Radioactive Wastes from the Nuclear Fuel Cycle, 2 Volumes

IAEA

Proceedings of a Symposium Organized by IAEA and OECD/NEA, Vienna, 1976, ISBN 92-0-020276-4

These volumes review the situation in the management of radioactive wastes generated by nuclear fuel facilities, to identify areas where important advances have been made, and to indicate areas where further technological development is needed. The proceedings include 62 papers by participants from 32 countries covering all aspects of managing nuclear fuel cycle wastes. The major topics of interest were high-level waste solidification techniques and disposal in geological formations.

KEY WORDS: waste management programs, solidification

A.9   Alternatives for Managing Wastes
  *   from Reactors and Post-Fission Operations in the LWR Fuel Cycle, 5 Volumes

ERDA

ERDA 76-43, 1976

These five volumes provide a comprehensive review of technical information expected to serve as a basis for decisions, assessments and environmental impact statements concerning the management of wastes from the back end of the commercial LWR fuel cycle. Volume 1 summarizes the report, describes alternatives, and provides background information on waste types and quantities. Volume 2 describes alternatives for waste treatment. Volume 3 discusses interim storage and transportation. Volume 4 describes alternatives for final storage and disposal. Volume 5, the appendixes, provides supplementary information with emphasis on characteristics of geologic formations that might be used as waste repository sites.

KEY WORDS: waste management programs, LWR

A.10   Ceramic and Glass Radioactive
        Waste Forms

        ERDA

        CONF-770102, January 1977

This report is a summmary of a meeting on
glass and polycrystalline ceramic radioactive waste forms, held at Germantown,
MD. Invidual presentations and group
reports on waste forms, material properties, and the behavior of glass and crystalline ceramics as waste forms are
included.

A.11   Behavior of Candidate Canister
        Materials in Deep Ocean Environment

        W. Smyrl and L. Stephenson, SLA

        SAND-76-9137; CONF-770303, March 1977

This paper, presented at a conference on corrosion (San Francisco, March 1977) describes
corrosion tests conducted under simulated
deep ocean conditions. Materials investigated were titanium, zirconium, and nickel
base alloys.

KEY WORDS: corrosion

A.12   Projections of Spent Fuel to be Dis-
        charged by the United States Nuclear
        Power Industry

        C. Alexander, C. Kee, A. Croff, and
        J. Blomeke, ORNL

        ORNL/TM-6008, October 1977

This report presents calculated properties
of spent fuel projected to be discharged
and accumulated by the U. S. nuclear power
industry though the year 2031. Extensive
detail is given and tables provided for
volume projections for many radionuclides.
Compilations are of grams of the element,
curies of radioactivity, thermal decay
power, photon and neutron emission rates,
and radiotoxicities of the assemblies
accumulated at a spent unreprocessed fuel
facility.

A.13   Plan for Solidification of Savannah
  *   River Plant High-Level Waste

        A. Jennings, SRL

        DP-MS-77-23; American Institute
        of Chemical Engineers 70th Annual
        Meeting, New York, N. Y., November
        1977

This report discusses plans for conversion
of the high-level wastes at Savannah River
Plant to high-integrity solids for transport to a federal repository. Details of
the process are given.

KEY WORD: solidification

A.14   Proceedings of the NEA/IAEA Tech-
        nical Seminar on Treatment, Condi-
        tioning, and Storage of Solid
        Alpha-Bearing Waste and Cladding
        Hulls, Paris, December 1977

        OECD and IAEA

        OECD, 1977

Sessions of the conference included discussions of general policies, with national program reviews from France, Germany, India, Japan, USSR, United Kingdom,
and the United States. Technical papers
addressed general management aspects of
solid alpha-bearing wastes, incineration
treatment, and conditioning concepts and
cladding hulls.

KEY WORDS: Germany, France, India,
                England, USSR, Japan

A.15   ORIGEN: Isotope Generation and
  *   Depletion Code-Matrix Exponential
        Method

        ORNL

        Radiation Shielding Information
        Center Computer Code Collection,
        CCC-217, 1977

This volume contains documentation and
sample problems for the ORNL code, ORIGEN,

a versatile point depletion code which solves the equation of radioactive growth and decay for large numbers of isotopes with arbitrary coupling. The complete text of ORNL-4628, with user's manual, is included in the package.

KEY WORDS: computer modeling

A.16   Alternatives for Long-Term Management
 *   of Defense High-Level Radioactive
        Waste - Savannah River Plant,
        2 Volumes

        ERDA

        ERDA 77-42, 1977

This report describes 23 alternative plans for long-term management of the high-level radioactive military wastes stored in tanks at the Savannah River Plant near Aiken, South Carolina. The description includes implementation technology, risk analyses, and cost estimates for the alternative plans for wastes generated at SRP through 1985. Thirty-one additional plans are developed for managing the waste that would result from continued SRP reactor operation through the year 1999. Cost estimates are given for these plans also.

KEY WORDS: waste management programs,
              military waste, high-level
              waste, risk analysis

A.17   Alternatives for Long-Term Manage-
 *   ment of Defense High-Level Radio-
        active Waste - Idaho Chemical
        Processing Plant

        ERDA

        ERDA 77-43, 1977

This report describes alternatives for long-term treatment and disposal of high-level radioactive military wastes stored as solids at the Idaho Chemical Processing Plant area of the Idaho National Engineering Laboratory. The existing waste management facilities and processes are described along with the

technology pertaining to various alternatives. Risks and costs are evaluated.

KEY WORDS: waste management programs,
              military waste, risk analy-
              sis, high-level waste

A.18   Alternative for Long-Term Man-
 *   agement of Defense High-Level
        Radioactive Waste, Hanford
        Reservation

        ERDA

        ERDA-77-44, 1977

This report describes alternatives considered for the long-term management of HLW accumulated as part of the national defense effort at the Hanford Reservation near Richland, Washington. Included are estimates of cost and biological risk and advantages, disadvantages, and status of technology for each alternative.

KEY WORDS: military waste, waste
              management programs

A.19   Techniques for the Solidifica-
        tion of High-Level Wastes

        IAEA

        ISBN 92-0-125077-0, 1977

This report is a review of the existing technology of high-level waste solidification, surveying and comparing all the work currently in progress. It examines the high-level liquid wastes arising from various processes under development or in operation, the advantages and disadvantages of each process for different types and quantities of waste solutions, the stages of development, the scale-up potential and flexibility of the processes.

KEY WORDS: high-level waste,
              solidification

A.20   Determination of Performance Criteria
for High-Level Solidified Nuclear
Waste

J. Cohen, LLNL

NUREG 0279, 1977

This study, prepared by LLL for the NRC, describes a system analysis model for the behavior of HLW solids for expected conditions and for postulated accidents as a function of the waste properties which affect radionuclide releases. Operational steps related to the management of HLW solids and included in this discussion are reprocessing plant operations, transportation, and pre- and post-emplacement phenomena at a repository. A format is given for specifying performance criteria directed to control of radionuclide release.

KEY WORDS: high-level waste

A.21   Effect of Internal Alpha Radiation on
Borosilicate Glass Containing Savannah
River Plant Waste

N. Bibler and J. Kelley, SRL

DP-1482, May 1978

Results are presented of the evaluation of the effects of internal alpha radiation on borosilicate glass for various samples, including a comparison sample of a glass containing Pu-238 without simulated waste. Glasses were examined for changes in physical stability, leachability, and dilation.

A.22   Inventory and Sources of Transuranic
 *   Solid Waste

TERA

UCRL-13934, August 1978

This report supports the development of standards and criteria which will specifically address the problem of transuranic (TRU) contaminated waste generated by Department of Energy (DOE) nuclear programs and commercial application of nuclear technology. The report presents an overview and introduction

to the generation, sources and past practices dealing with TRU-contaminated waste. There are also sections on DOE facilities, commercial disposal sites, a waste inventory, waste projections; and an extensive list of references and related bibliography are given.

KEY WORDS: transuranic waste

A.23   Waste Acceptance Criteria Study -
Phase I: Identification of
Characteristics

ONWI

ONWI-6(1), September 1978

Waste management operations affecting the characteristics of several waste types are discussed in terms of criteria required for a waste management program. Nuclear, chemical, physical/mechanical, and thermal characteristics are identified for low-level and intermediate-level transuranic high-level, spent-fuel, cladding, fission product, and contaminated equipment wastes. The operations considered are generation, treatment, packaging, transportation, interim storage, short-term disposal, long-term disposal, and retrieval.

KEY WORDS: waste management programs

A.24   Proceedings of the Conference on
 *   High-Level Radioactive Solid
Waste Forms, Denver, CO,
December 1978

L. Casey, ed., NRC

NUREG/CP-0005, 1978

This set of papers includes discussions of 1) vitreous forms, 2) encapsulation techniques and failure modes, 3) spent fuel, crystalline, and other forms, and 4) assessment of the DOE program for high-level waste immobilization. Presentations are focused on developments and applications of materials science for high-level radioactive waste disposal, including solidification of wastes, waste canisters, and potential buffering overpacks. Authors and papers represent national and international programs.

KEY WORDS: solidification, waste management programs, Germany, Sweden, France

A.25   Radioactive Wastes at the Hanford
*   Reservation - A Technical Review

        National Research Council, Washington, D. C.

        National Academy of Sciences, 1978

This report is a review and evaluation of the management of military radioactive wastes at the Hanford reservation in southeastern Washington. The report includes a description of the amounts and condition of various kinds of wastes, a critique of practices and plans for managing the wastes, and a review of plans for long-term waste disposal. Emphasis is on adequacy of technical management practices to ensure protection of Hanford personnel and the surrounding population from radiation hazards. The report also suggests areas for research.

KEY WORDS: waste management programs, military waste

A.26   Proceedings of the Symposium on "Science Underlying Radioactive Waste Management," Materials Research Society, Boston, MA, 1978. Scientific Basis for Nuclear Waste Management, Volumes 1 and 2

*   G. McCarthy, ed., Vol. 1 Materials Research Laboratory, Pennsylvania State University, University Park, PA

*   C. Northrup, ed., Vol. 2, SLA

These proceedings include papers covering the range of basic and applied sciences that form the basis for nuclear waste management. The general headings are: waste solidification, waste isolation, waste treatment, and modeling and safety assessment.

KEY WORDS: solidification, migration, risk analysis, geologic properties

A.27   Proceedings of the Symposium on the On-Site Management of Power Reactor Wastes, Zurich, March 1979

        OECD, IAEA, and the Swiss Office for Science and Research

        OECD, 1979

Sessions are devoted to radioactive waste management practices at nuclear power plants, waste arisings and operating experiences, coolant and liquid waste treatment, solidification methods, volume reduction methods, and solid waste containment.

A.28   Characteristics of Defense High-
*   Level Waste

        H. Cheung and D. Kvam, LLNL, and B. Knaizewycz, TERA

        NUREG/CR-0685, UCRL-52704, May 31, 1979

This report presents a historical and technical review of the high-level radioactive waste management problems at Savannah River Plant, Hanford Reservation tion, and Idaho Chemical Processing Plant with major emphasis on high-level waste inventory, characteristics, and management technologies. This information is presented for use as a portion of the defense high-level waste database being developed to provide the U. S. Nuclear Regulatory Commission with the necessary information to develop its regulatory framework and guidelines.

KEY WORDS: military waste, database

A.29   Assumptions and Ground Rules Used in Nuclear Waste Projections and Source Term Data

        S. Storch and B. Prince, Union Carbide Corporation, Oak Ridge, TN

        ONWI-24, September 1979

Assumptions and ground rules of long-term domestic commercial nuclear waste projections are compared for four studies: Union Carbide Office of Waste Isolation Study; Arthur D. Little, Inc. Study; DOE Study; and Battelle Pacific Northwest Laboratory Commercial Waste Management Impact Statement. The studies are compared with respect to waste form characteristics, packaging, and shipment. Issues related to interim storage are discussed. Assumptions and limitations of specific computer codes in the studies are outlined.

KEY WORDS: computer modeling, waste management programs

A.30   Spent Fuel Database: Commercial Light Water Reactors

TERA

UCRL-15186, December 1979

A comprehensive database describing light water reactor (LWR) fuel technology is compiled. This document provides the technology baseline and supports the development of evaluation standards and criteria applicable to the disposal of spent nuclear fuel. Four categories of information are included in the spent fuel database: 1) physical characteristics of the spent fuel, 2) composition of the spent fuel and fuel bundles in the spent fuel pool, 3) spent fuel transportation and shipping cask descriptions, and 4) accident considerations and potential radioactive releases.

KEY WORDS: LWR, database

A.31   An Assessment of LWR Spent Fuel Disposal Options (3 Volumes)

Bechtel National, Inc., San Francisco, CA

ONWI-39, 1979

This study, part of DOE's NWTS Program, is an evaluation of three specific ultimate disposal methods for spent nuclear fuel assemblies. Alternative forms of

spent fuel are identified and the physical systems and facilities for processing, handling and transporting each of them are defined. The alternatives are analyzed and rated in terms of several assessment criteria: technical feasibility, safeguards, criticality, radiological impact, retrievability, and economics.

A.32   Long-Term High-Level Waste Technology Program, Strategy Document

SRL

DOE/SR-WM 79-3, April 1980

This strategy document presents the research and development plan of DOE/IS Division of Waste Products for long-term immobilization of the high-level radioactive wastes resulting from chemical processing of nuclear reactor fuels and targets. Several options are presented, with site-specific processes for immobilization to tailor waste forms to candidate geologic repositories.

KEY WORDS: waste management programs

A.33   Proceedings of a Workshop on Alternative Nuclear Waste Forms and Interactions in Geologic Media, Gatlinburg, TN, April 1981

ORNL and DOE

CONF-8005107

The papers in this collection address the conference purposes: to review the status of the research on alternate waste forms, to compare characteristics of alternate waste forms with those of glasses, and to encourage interaction between basic research and applied aspects of the waste form issue. The papers include discussions on leaching, radiation damage, and hydrothermal interaction of ceramics with basalt, and a review of heat dissipation in geologic media.

A.34   Solidification of High-Level
*   Radioactive Wastes

       Committee on Radioactive Management
       National Research Council

       National Academy of Sciences

This extensive report, prepared by a panel on
waste solidification, includes 1) analysis of
the role that properties of various solid
forms play in determining selection of the
form appropriate to a particular waste management program, 2) information on research
into waste forms, 3) evaluation of management
of R&D for waste solidification, and 4) recommendations to federal agencies concerning
solid form options.

KEY WORDS: solidification

A.35   Immobilization of High-Level Nuclear
       Reactor Wastes in Synroc

       A. Ringwood, S. Kesson, N. Ware,
       W. Hibberson, and A. Major,
       Research School of Earth Sciences,
       Australian National University,

This paper presents the arguments in favor
of the immobilization of radioactive waste
in Synroc prior to its disposal in a geological medium.   The report describes the
technique of incorporating waste in Synroc
materials, leaching tests on Synroc and
glasses and the operation of the Synroc
process.

KEY WORDS: Synroc

A.36   Handling of Spent Nuclear Fuel and
       Final Storage of Vitrified High-
       Level Reprocessing Waste (5 Vols.)

       KBS

       1977

       (SEE ABSTRACT C.61.)

A.37   Modelling Studies Used to
       Evaluate Waste Disposal Options

       P. Grimwood, M. Hill, and G. Webb,
       NRPB, England

       Nuclear Engineering International
       23, January 1978, p. 55

       (SEE ABSTRACT C.65.)

A.38   Report to the President

       IRG

       TID-28817 (Draft), 1978

       (SEE ABSTRACT C.100.)

A.39   Handling and Final Storage
       of Unreprocessed Spent Nuclear
       Fuel (2 Vols.)

       KBS

       1978

       (SEE ABSTRACT C.104.)

A.40   Alternative Waste Disposal Con-
       cepts - An Interim Technical
       Assessment

       D. Crandall, Bechtel National,
       Inc., San Francisco, CA

       ONWI-65, November 1979

       (SEE ABSTRACT C.152.)

A.41   Alternative Processes for Managing
       Existing Commercial High-Level
       Radioactive Wastes

       BPNL

       NUREG-0043

       (SEE ABSTRACT C.194.)

A.42   Survey of Naturally Occurring Ha-
       zardous Materials in Deep Geologic
       Formations: A Perspective on the
       Relative Hazard of Deep Burial of
       Nuclear Wastes

       K. Tonnessen and J. Cohen

       UCRL-52199, January 14, 1977

       (SEE ABSTRACT D.9.)

A.43   An Analysis of Capital and
       Operating Costs Associated with
       High-Level Waste Solidification
       Processes

         B. Kniazewycz, TERA and
         R. Heckman, LLNL

         UCRL-80064, March 1978

       (SEE ABSTRACT D.16.)

A.44   Accident Risk Assessment -
       Status Report on the EPRI
       Fuel Cycle

         Science Applications, Inc.
         Palo Alto, CA

         EPRI NP-1128, 1979

       (SEE ABSTRACT D.25.)

A.45   State of Waste Disposal Tech-
       nology and The Social and
       Political Implications

         R. Post, ed., University of Arizona

         Proceedings of the Symposium on
         Waste Symposium on Waste Manage-
         ment, sponsored by the University
         of Arizona College of Engineers
         and the Arizona Energy Commission,
         Tucson, Arizona, 1979

       (SEE ABSTRACT D.26.)

B.   TRANSPORTATION OF WASTE

B.1   Severities of Transportation
*   Accidents, 4 Volumes

R. Clarke, J. Foley, W. Hartman,
and D. Larson, SLA

SLA-74-0001, July 1976

These volumes provide quantitative descriptions of the severities of the environments that small containers of hazardous materials can be expected to experience in transportation accidents. A comprehensive outline of the data and methodology are given. In Volume I, significant results of the study are summarized; in Volume II, physical parameters of aircraft accident environments are discussed; in Volumes III and IV, truck and train accidents are examined.

B.2   An Overview of Transportation in
the Nuclear Fuel Cycle

R. Rhoads, BPNL

BNWL-2066, May 1977

A review is given of current transportation systems and regulations for radioactive materials in the nuclear fuel cycle. The report includes projections of the growth of nuclear power, estimated annual shipments of full cycle materials, and the projected transportation requirements of the nuclear power industry through the year 2000.

B.3   Developing Criteria for the Management of Nuclear Wastes

R. Heckman, LLNL

UCRL 52000, Energy & Technology
Review, October 1977

This study reports on the development of the mathematics for comprehensive and systematic modeling of complete nuclear waste disposal scenarios. For the widely accepted waste disposal scheme of deep geologic burial, research here suggests transportation to the disposal site as potentially the most hazardous aspect. The work indicates several performance criteria that would greatly decrease the risk.

KEY WORDS: computer modeling,
risk analysis

B.4   Environmental Statement on the
*   Transportation of Radioactive
Material by Air and Other Modes,
2 Volumes

NRC

NUREG-0170, 1977

This statement was prepared in connection with NRC's reevaluatin of regulations governng air transportation of radioactive materials to provide analysis for decisions concerning the effectiveness of existing

27

rules and possible alternatives to the rules. As alternatives, other modes of transportation are examined with respect to the effect on packaging as related to the exposure of people under both normal and accident situations. Security and safeguards are considered, with an assessment of physical security requirements applied to each mode of transportation.

KEY WORDS: radiological consequences, regulations, safeguards, risk analysis, dose, environment

B.5   Everything You Always Wanted to Know About Shipping High-Level Nuclear Waste

        DOE

        DOE/EV-0003, WASH-1264, Revision January 1978

This is a collection of most often asked questions about the safety of shipping high-level nuclear wastes. Detailed answers to the questions are provided in an effort to inform the public. The composite parts of the overall transportation problem are presented. Sample questions: How will high-level waste be transported? Will the waste shipments be routed around cities and congested areas? Who inspects shipments and shippers?

B.6   Radioactive Waste Transportation Systems Analysis and Program Plan

        L. Shappert, D. Joy, and M. Heiskell, ORNL

        ORNL-5362, March 1978

This report describes a model for determining optimum shipping policies for low-level transuranic waste from generating sites to federal respositories. The overall systems analysis program developed to identify and formulate plans to deal with waste transportation problem is presented.

KEY WORDS: low-level waste, transuranic waste

B.7   Proceedings of the Fifth International Symposium on Packaging and Transportation of Radioactive Materials, Las Vegas, May 1978, Volumes I and II

The symposium was organized to review the state of the art of the technology and politics of radioactive material transportation and to stimulate discussion towards solutions of the problems involved. The session topics include transportation systems, sea transport, rail transport, transportation and packaging systems for plutonium, and heat transfer. Technical papers and the keynote address are presented in these volumes.

B.8   Alternatives for Managing Wastes from Reactors and Post-Fission Operations in the LWR Fuel Cycle, 5 Volumes

        ERDA

        ERDA 76-43, 1976

        (SEE ABSTRACT A.9.)

B.9   Alternatives for Long-Term Management of Defense High-Level Radioactive Waste - Savannah River Plant, 2 Volumes

        ERDA

        ERDA 77-42, 1977

        (SEE ABSTRACT A.16.)

B.10   Determination of Performance Criteria for High-Level Solidified Nuclear Waste

        J. Cohen, LLNL

        NUREG 0279, 1977

        (SEE ABSTRACT A.20.)

B.11   An Assessment of LWR Spent Fuel
       Disposal Options (3 Volumes)

       Bechtel National, Inc.,
       San Francisco, CA

       ONWI-39, 1979

       (SEE ABSTRACT A.31.)

B.12   **Handling of Spent Nuclear Fuel
       and Final Storage of Vitrified
       High-Level Reprocessing Waste
       (5 Vols.)**

       KBS

       1977

       (SEE ABSTRACT C.61.)

B.13   Report to the President

       IRG

       TID-28817 (Draft), 1978

       (SEE ABSTRACT C.100.)

B.14   Handling and Final Storage
       of Unreprocessed Spent Nuclear
       Fuel (2 Vols.)

       KBS

       1978

       (SEE ABSTRACT C.104.)

B.15   Alternative Waste Disposal
       Concepts - An Interim Technical
       Assessment

       D. Crandall, Bechtel National,
       Inc., San Francisco, CA

       ONWI-65, November 1979

       (SEE ABSTRACT C.152.)

B.16   Management of Radioactive Waste
       Gases from the Nuclear Fuel
       Cycle - Volume I: Comparison
       of Alternatives

       A. Evans, W. Prout, J. Buckner,
       M. Buckner, SRL

       NUREG/CR-1546, December 1980

       (SEE ABSTRACT C.183.)

B.17   Alternative Processes for Man-
       aging Existing Commercial High-
       Level Radioactive Wastes

       BPNL

       NUREG-0043

       (SEE ABSTRACT C.194.)

B.18   Accident Risk Assessment -
       Status Report on the EPRI
       Fuel Cycle

       Science Applications, Inc.,
       Palo Alto, CA

       EPRI NP-1128, 1979

       (SEE ABSTRACT D.25.)

B.19   State of Waste Disposal Tech-
       nology and The Social and
       Political Implications

       R. Post, ed., University of
       Arizona, Tucson, Arizona

       Proceedings of the Symposium
       on Waste Management, sponsored
       by the University of Arizona
       College of Engineers and the
       Arizona Energy Commission,
       Tucson, Arizona, 1979

       (SEE ABSTRACT D.26.)

C. SITE SELECTION AND REPOSITORY DESIGN

C.1 The Behavior of Radioactive Fission Products in Soils

V. Klechkovsky and G. Tselishcheva, Academy of Sciences of the U.S.S.R. Moscow

AEC-TR-2867, On the Behavior of Radioactive Fission Products in Soil, Their absorption by Plants and Their Accumulation in Crops, February 1957

Sorption and elution studies used to examine the behavior of fission products in soil are described. Experimental tests run using Sr 90, Sr 89, Y 90, Cs 137, Zr 95, Nb 95, Ru 106 and Rh 106 are analyzed.

C.2 The Disposal of Radioactive Waste
   * on Land

NAS-NRC Pub. No. 519, September 1957

This report presents the proceedings of a conference on radioactive waste disposal. Four geologic categories are discussed: surface excavation, infiltration into shallow permeable beds, natural and artificial excavations, and artificial solution cavities. Recommendations for further site investigation are made.

C.3 Proceedings of an International Colloquium on Retention and Migration of Radioactive Ions in Soils, Saclay, France, October 1962

ORNL-TR-4638, 1962

This collection of papers includes reports on underground storage conditions for nuclear wastes, and other technical topics related to radionuclide transport.

KEY WORDS: brine migration, transport

C.4 Montmorillonite Exchange Equilibria with Strontium-Sodium-Cesium

J. Eliason, BPNL

American Mineralogist $\underline{51}$, March-April, 1966, p. $\overline{324}$

This report describes experimental studies on the Bayard, New Mexico, montmorillonite. Ion exchange isotherms and the free energy changes of -2278, -331, and -4019 cal/mole for the cesium-sodium, strontium-sodium, and cesium-strontium systems, respectively, were determined. Selectivities across ranges of loading and some results were compared with those from studies on the Chambers, Arizona, montmorillonite.

C.5 The Long-Term Hazard of Radioactive Wastes Produced by the Enriched Uranium, Pu-^{238}U, and ^{233}U-Th Fuel Cycles

M. Bell and R. Dillon, ORNL

ORNL-TM-3548, November 1971

An evaluation is presented of the long-term hazards of high-level and alpha radioactive wastes generated by irradiation of three fuels representative of those which will be used to generate electric power in the next decades. The composition and radioactivity of the wastes generated by typical LWR, LMFBR and molten-salt breeder reactors using the enriched uranium, Pu-^{238}U, and ^{233}U-Th fuel cycles, respectively, have been computed for times up to 30 million years after discharge. The hazard of various types of wastes contaminated to 10 µCi/kg of initial parent alpha activity was also calculated and the time of maximum relative hazard determined.

KEY WORDS: radiological consequences

C.6   Considerations in the Long-Term
       Management of High-Level Radioactive
       Wastes

       F. Gera and D. Jacobs, ORNL

       ORNL-4762, February 1972

Projections are made of the amounts of
radioactive wastes accumulated to 2020.
Significant waste radionuclides are compared
on the basis of their potential hazard to man
over their physical lifetimes. Waste solidification and deep geologic disposal are presented as the optimal waste management program.

KEY WORDS: radiological consequences,
            waste management programs

C.7   Transportation of Radioactive Waste
       Materials to the Sun

       R. Reichert, Dornier System Gmbh,
       Germany

       CONF-721068; Proceedings of the 23rd
       Congress Symposium on Nuclear Power
       and Propulsion Devices in Space,
       Vienna, Austria, October 1972

Initial results are given for a feasibility
study of applying space technology for the
removal of radioactive wastes to the sun.
Technical problems, such as trajectory
options, launch vehicles, and transport costs
are discussed.

C.8   Neutron-Induced Transmutation of
       High-Level Radioactive Waste

       H. Claiborne, ORNL

       ORNL-TM-3964, December 1972

The possibility of reducing the potential
hazard of high-level radioactive waste by
neutron-induced transmutation is examined.
This report reviews and discusses the available information on fission product transmutation and provides calculations of the contribution of individual actinides to the potential hazard of waste. Expected hazard
reduction factors that would result from

recycling through a PWR are calculated for
the actinide waste from chemical processing of spent fuel.

KEY WORDS: radiological consequences,
            actinides

C.9   Putting Radioactive Waste in Ice -
       A Proposal for an International
       Radionuclide Depository in Antarctica

       E. Zeller, D. Saunders, and E. Angino,
       University of Kansas, Lawrence, KS
       and Texas Instruments, Dallas, TX

       Bulletin of the Atomic Scientists
       29, January 1973, p. 4

This report describes the advantages of
sites in Antarctica for permanent disposal
of international nuclear waste. Recommendations are made for further investigation into issues such as ice-canister
interactions, ice movements, sub-ice topography, and temperature gradients.

C.10   Storage Fee Analysis for a Retriev-
       able Surface Storage Facility

       B. Field and C. Rosnick, Atlantic
       Richfield Hanford Company,
       Richland, WA

       ARH-2960, December 1973

This report is a revision of an earlier
report (ARH-2746) and is concerned with
establishing a fee to cover the cost of
storing nuclear waste in a retrievable
surface storage facility and in a subsequent disposal facility. A broader and
more flexible model than the earlier model
for determining the fee is presented.

KEY WORD: retrievability

C.11   Feasibility of Space Disposal
       of Radioactive Nuclear Waste

       National Aeronautics and Space
       Administration, Cleveland, Ohio

       NASA TM X-2911, December 1973

An investigation is reported on the feasibility of disposing of nuclear waste in

space. This report presents tentative solutions for technical problems of safely packaging the separated long-lived actinide wastes. A cost analysis is given along with recommendations for a follow-up experimental program and safety assessment to establish a system design.

KEY WORDS: waste management programs

C.12   High-Level Radioactive Waste
        Management Alternatives

        AEC

        WASH-1297, May 1974

This study summarizes information on potential alternative methods for long-term management of high-level radioactive wastes. Issues considered include technical feasibility, safety, cost, the environment, policy conflicts, public response, and research and development needs for waste disposal in terrestrial locations, in space, and by elimination (nuclear transformation).

KEY WORDS: waste management programs
              high-level waste, environment

C.13   Retrievable Surface Storage
        Facility Alternative Concepts
        Engineering Studies

        Atlantic Richfield Hanford Company
        Richland, WA

        ARH-2888, Revision, July 1974

General design criteria are described for a retrievable surface storage facility capable of receiving high-level radioactive wastes generated by commercial reactor fuel reprocessing plants through the year 2000 and storing these wastes under surveillance. The concepts studied are 1) storage in water basins where the decay heat is rejected to the atmosphere by the use of circulating cool water, heat exchangers, and cooling towers, 2) storage in air-cooled vaults where heat removal is by natural convection, and 3) storage in casks placed outdoors and cooled by convective air flow. The report presents engineering features and an engineering basis for safety, environmental, and cost analysis for each concept.

KEY WORDS: retrievability, waste

management programs

C.14   Environmental Statement Management
        of Commercial High-Level and Tran-
        uranium-Contaminated Radioactive
        Waste

        AEC

        WASH-1539, Draft, September 1974

This report is the AEC's environmental statement on the consequences of 1) developing an engineered surface storage facility for retrievable storage of commercial high-level waste, 2) evaluating geological formations as potential repository sites, and 3) providing retrievable storage for commercial transuranic-contaminated waste pending availability of permanent disposal. The document includes background information, possible waste management methods, and cost-benefit analyses of alternatives. This statement is replaced by DOE/EIS-0046-D, April 1979 (see C.126).

KEY WORDS: waste management programs,
              environment

C.15   Rock Types, Also Geologic and Hydro-
        logic Settings, Favorable to Deep
        Placement of High-Level Radioactive
        Wastes

        Y/OWI/SUB-3745/4, October 1974

This report identifies those rock types and geologic-hydrologic settings in which a high-level waste repository might be expected to remain functionally intact for at least 1000 years, and possibly hundreds of thousands of years, deeply buried underground. Rock characteristics are explored and criteria presented for site suitability.

KEY WORDS: rock, salt, shale

C.16   Long-Term α-Hazard of High Activity
        Waste from Nuclear Fuel Reprocessing

        F. Girardi and C. Bertozzi, CEC

        EUR 5214e, 1974

This is a summary of a study on the long-term hazard of α-emitting radionuclides and of the experimental activity on the feasibility of the chemical separation of actinides and their successive burn-out in a reactor. Relative merits of various decontamination processes and the associated uncertainties are discussed briefly.

KEY WORD: actinides

C.17   Conclusions Drawn from Studies on the Oklo Natural Nuclear Reactor Concerning the Evolution in the Ground of Radioactive Waste Over a Long Period

C. Frejacques, Commission of Atomic Energy, France

CONF-750411; Proceedings of the First European Nuclear Society Conference, Paris, France, April 1975

This paper presents preliminary results of a localized study of isotopes present at the site of the Oklo natural reactor. Rare earths, Pu, Kr and Xe, are traced and conclusions about relative hazards are drawn.

KEY WORD: transport

C.18   Proceedings of the First European Nuclear Society Conference on Nuclear Energy Maturity, Paris, France, April 1975

American Nuclear Society, 1975

Papers at this conference address the general problems of nuclear energy and several nuclear waste management programs in particular. Technical issues, such as the interaction between processing methods and burial sites, are reviewed.

C.19   Feasibility of Sealing Boreholes with Compacted Natural Earthen Material, 3 Volumes

MIT

ORNL/SUB-3960/2, June 1975

This research report describes laboratory investigations of borehole sealing using compacted shale or shale component materials in a shale substrate. Materials were examined for permeability, strength, shrinkage, and swell pressure. Volume 2

contains detailed appendices of work carried out in Volume 1. Volume 3, issued in June 1976, continues the investigation.

KEY WORDS: shale

C.20   Compilation of Hanford Corrosion Studies

D. Lini, Atlantic Richfield Hanford Company, Richland, WA

ARH-ST-111, July 1975

This is a review of experimental corrosion studies conducted at Hanford over 30 years, identifying types and rates of corrosion, describing data-gathering techniques, and identifying the contribution of the data to tank failure predictions. Studies include general and pitting corrosion, stress corrosion cracking, corrosivity of solidified high-level waste, and cathodic protection.

KEY WORDS: corrosion

C.21   Disposal of Radioactive Wastes in Salt Domes

F. Gera, ORNL

Health Physics, $\underline{29}$, July 1975
p. 1

Questions concerning the use of diapiric salt structures as media for radioactive waste disposal are investigated. A summary of technical problems, associated with tectonic stability, salt dissolution, and reliability of long-term containment is given. Comparison is made between the emplacement concepts of 1) mechanical mine and 2) the mixing of granular waste and crushed salt in deep solution cavities with respect to technical problems.

KEY WORDS: salt

C.22   Geochemical Behavior of Long-Lived Radioactive Wastes

F. Gera, ORNL

ORNL-TM/4481, July 1975

This report examines evidence of reduced hazard potential over time of radioactive waste. Data presented includes that on the levels of activity in foods grown on radioactive soils and surveys of mill tailings.

C.23 Hydrology of Some Deep Mines in Precambrian Rocks

D. Yardley, University of Minnesota, Minneapolis, MN

Y/OWI/SUB-4367/1, October 1975

Studies on several deep mines in Precambrian rocks of the Lake Superior region are reported. Details of mine geology, hydrology and structural and stratigraphic sections are provided.

KEY WORDS: geologic properties, hydrologic properties

C.24 Are Plutons the Answer to Nuclear Waste Disposal?

J. Carruthers, F. P. Publications, Ottawa, Ontario, Canada

Science Forum 8, (6), December 1975, p. 15

This article is a review of Atomic Energy of Canada Limited's policy with respect to storage of high-level radioactive wastes and the initiation of a joint investigation with the Geologic Survey of Canada to study salt deposits and granitic plutons as possible sites for long-term, permanent, underground storage.

KEY WORDS: Canada, waste management programs, granite

C.25 Radiolysis and Temperature Effects in Case of Underground Storage of Bitumen

E. Smailos, W. Diefenbacher, E. Korthaus, and W. Comper, Gesellschaft für Kernforschung mbh, Karlsruhe, Germany

KFK-2329, May 1976

This report attempts to define limits to the specific activity of bituminized wastes

stored in a prototype cavity which will avoid buildup of combustible radiolytic gases and overheating of stored waste. Details of calculations leading to recommendations for limits on activity for safe disposal of wastes are provided.

C.26 Proceedings of an International Symposium on Management of Waste from the LWR Fuel Cycle, Denver, CO, July 1976

CONF-760701, 1976

Papers in this conference include reviews of waste management programs in the U. S. and in foreign countries as well as technical papers in waste isolation and related issues. Sessions are on governmental responsibilities and the public's interest; the nature of wastes and options for waste management; solid, liquid, and gaseous wastes; packaging and transport; pathways and exposure potentials; and isolation in geologid formation.

KEY WORDS: waste management programs

C.27 Terminal Storage of Radioactive Wastes in Geologic Formations

T. Lomenick, OWI

CONF-760744; Proceedings of a Symposium on Energy Sources for the Future, Oak Ridge, TN, July 1976

This paper reviews the goals and studies associated with the NWTS program. Further studies on shale and clay in the U. S. are recommended. Investigations of several areas of argillaceous rock are reviewed.

C.28 A Critical Study on the IAEA Definition of High-Level Radioactive Waste Unsuitable for Dumping at Sea

T. Miyake and K. Saruhashi, Geochemistry Research Association, Tokyo, Japan, and Meteorological Research Institute, Tokyo, Japan

Papers in Meteorology and Geophysics, 27 (3), September 1976, p. 75

This is a critical review of the definition of high-level waste unsuitable for

sea disposal set forth by the London Conference on the Dumping of Wastes at Sea, 1972,
and adopted by the IAEA in 1975. A reexami-
nation by IAEA is recommended.

KEY WORD: environment

C.29.   Evaluation on the Disposal of Radio-
        active Waste into the North Pacific

        Y. Sugiura, K. Saruhashi, and
        Y. Miyake, Meterological Research
        Institute, Tokyo, Japan, and Geo-
        chemistry Research Association,
        Tokyo, Japan

        Papers in Meterorology and Geophysics,
        27, September 1976, p. 81

Evaluation of deep sea disposal of radioactive waste into the North Pacific Ocean
on the basis of diffusion processes is presented using a simplified model. Calculations are made for release and dose evaluation.

C.30   Waste Isolation Facility Description:
        Bedded Salt

        Parsons, Brinckerhoff, Quade and
        Douglas, Inc., New York

        Y/OWI/SUB-76/16506, September 1976

This report contains a detailed description
of the pilot waste isolation facility, including shaft design and characteristics,
design and construction specifications, design criteria, regulations and codes, and
layout drawings.

KEY WORDS: bedded salt

C.31   Assessment of the Radiological Pro-
        tection Aspects of Disposal of High-
        Level Waste on the Ocean Floor

        P. Grimwood and G. Webb, National
        Radiological Protection Board,
        England

        NRPB-R-48, October 1976

This study is a preliminary assessment of
the potential radiological consequences of
disposal of solidified high-level radioactive
waste on the floor of the deep ocean. The

report describes modeling of radionuclide
release into water, its dispersion in the
ocean and eventual uptake in marine organisms and sediments. Radiological consequences are assessed. Areas for future
study are identified.

KEY WORDS: radiological consequences

C.32   Radiolytic Gas Generation in Plu-
        tonium Contaminated Waste Materials

        A. Kazanjian, Rockwell International
        tional, Golden, CO

        RFP-2469, October 1976

Studies are reported on determinations of
the extent of hazards crested by plutonium
contaminated waste materials decomposing
into gaseous products. Gas generation
yields were measured for common waste materials contacted with plutonium and results
are given.

KEY WORD: plutonium

C.33   The Response of Salt Migration
        to Dissolution in Salt Domes and
        How It Affects a Radioactive
        Waste Isolation Facility Design

        J. Hamstra, Netherlands Energy
        Research Foundation, Petten N. H.,
        The Netherlands

        CONF-76115; Proceedings of a
        Symposium on Salt Dome Utiliza-
        tion and Environmental Considerations
        Baton Rouge, LA, November 1976

This paper investigates a potential failure mechanism for long-term isolation of
radioactive wastes: that groundwater
will contact the rock salt containment
shield so that a continuous dissolving
action could take place and result in
radionuclide migration into the biosphere.
Evidence is provided to suggest that a
salt dome in contact with its mother salt
layer will, under certain conditions, res-
pond to a continuous groundwater attack by
a salt migration towards the surface area
where the dissolving action takes place.
model calculations are given and results
discussed.

KEY WORDS: brine migration

C.34   Proceedings of the 24th Conference
        on Remote Systems Technology,
        Washington, D. C., November 1976

        R. Farmakes, ANS, ed.

These proceedings of a conference held in
conjunction with the American Nuclear Society
include technical papers on 1) aspects of the
behavior and handling of tritium in nuclear
facilities, 2) waste management equipment and
processes, 3) impact of ALARA on present and
future nuclear facilities, and 4) advances in
remote manipulation.

C.35   Management of Radioactive Wastes

        W. Lennemann, H. Parker, and P. West,
        IAEA

        Annals of Nuclear Energy 3, 1976,
        p. 285

This article addresses concerns of nuclear
waste management: management and control
over mill tailings, a shortage of irradiated
fuel storage space, and decommissioning criteria. International cooperation in resolving waste disposal questions and minimizing
the number of facilities involved in fuel
reprocessing and radioactive waste management
is recommended.

KEY WORDS: waste management programs,
            decommissioning

C.36   Antarctica - A Potential International
        Burial Area for High-Level Radioactive
        Wastes

        E. Angino, G. Dreschhoff and E. Zeller,
        University of Kansas, Lawrence, KS

        Bulletin of the International Asso-
        ciation of Engineering Geology, 14,
        1976, p. 173

Disposal of solid high-level radioactive
wastes in Antarctica is considered through
two separate approaches: burial in the
massive ice sheet of East Antarctica and
burial in the bedrock in the dry valley or
oasis areas of East Antarctica. Summaries
of advantages of and objections to these
disposal methods are given.

C.37   Antarctica, A Potential Disposal
        Site for the World's High-Level
        Radioactive Waste

        E. Zeller, E. Angino, and D. Saunders,
        University of Kansas, Lawrence, KS
        and Texas Instruments, Dallas, TX

        Modern Geology, 6, 1976, p. 31

This report discusses the potential of large
masses of ice for the disposal of radioactive wastes. A summary of work on fractures, heat conductivity, and stability as
factors of concern in site selection is
given.

C.38   High-Level Radioactive Wastes
        from Light-Water Reactors

        B. Cohen, University of
        Pittsburgh, PA

        Reviews of Modern Physics, Vol. 49,
        No. 1, January 1977

This article traces the production of radioactive nuclei during the operation of a
LWR and follows their decay history. The
potential environmental impacts of this
waste are calculated and compared to other
materials.

KEY WORD: LWR

C.39   Proceedings of an International Con-
        ference on Nuclear Power and Its
        Fuel Cycle, Salzburg, Austria,
        May 1977

        IAEA, 1977

Papers presented cover all aspects of the
nuclear fuel cycle, including discussions
of waste form and candidate site suitability.

C.40   Proceedings of a Workshop on Risk
        Analysis and Geologic Modeling in
        Relation to the Disposal of Radio-
        active Wastes into Geological For-
        mations, Italy, May 1977

        OECD and CEC

        May 1977

Papers are included from sessions summarizing results of risk analysis of repositories for radioactive wastes in geologic salt formations in foreign countries, and from sessions on geosphere transport, containment failure modes, and safety assessment methodology.

KEY WORDS: risk analysis

C.41   Retention of Plutonium and
        Americium by Rock

        S. Fried, A. Friedman, R. Atcher,
        and J. Hines, Argonne National
        Laboratory, Argonne, IL

        Science 196, June 3, 1977, p. 1087

This is a summary of experiments to determine migration and retention of the radionuclides Pu 237, Pu 238 and Am 241 in selected rock types. The migratory behavior of actinides away from radioactive waste disposal sites is inferred and extrapolations made which allow for catastrophic intrusion of water or seismic activity resulting in fissures. Conclusions are drawn about repository design and site suitability.

KEY WORDS: transport, plutonium

C.42   Oceanic Distribution of Radio-
        nuclides from Nuclear Weapons
        Testing

        V. Bowen, Woods Hole Oceanographic
        Institute, Woods Hole, MA

        CONF-770611; Transactions of the
        American Nuclear Society, Annual
        Meeting, New York, N. Y., June 1977

This report reviews information derived from movement in the oceans of atmospheric fallout from nuclear weapons testing, as applicable to studies of seabed disposal of radioactive wastes. Parameters such as density, rates of sinking, fixation rates, and velocity are investigated.

KEY WORD: transport

C.43   United States Seabed High-Level
        Radioactive Waste Disposal Feasi-
        bility Study

        D. Talbert and D. Anderson, SLA

CONF-771109; Transactions of the American Nuclear Society, Annual Meeting, New York, N. Y., June 1977

This paper reports on a multidisciplinary investigation of the feasibility of high-level waste disposal in submarine geologic formations. The program phases of 1) data gathering, 2) the development of the barrier concept, and 3) the investigation of sediments, chemistry, and thermal and radiation environments that might affect emplacement and containment within the sea floor are summarized.

KEY WORD: transport

C.44   Preliminary Geologic Site Selection
        Factors for the National Waste
        Terminal Storage Program

        Woodward-Clyde Consultants,
        San Francisco, CA

        Y/OWI/SUB-76/16531, June 1977

Geologic considerations, including depth, volume, tectonic stability, and hydrologic regime of host rocks, are described for a radioactive waste repository. A description of information necessary to assess geologic factors and to determine the acceptability of candidate sites is presented. Charts and tables of geologic siting objectives and criteria are also included.

KEY WORDS: geologic properties,
            hydrologic properties, rock

C.45   Underground Disposal of Canada's
        Nuclear Waste

        H. Tammenagi, J. Gale, and
        B. Sanford, Atomic Energy of
        Canada Limited, Whiteshell
        Nuclear Research Establishment,
        Pinawa, Manitoba, Canada and
        Geological Survey of Canada,
        Ottawa, Ontario, Canada

        Geoscience Canada, 4 (2),
        June 1977, p. 71

This is an outline of Canadian programs in process and planned for the future to ensure the long-term underground isolation of radioactive wastes. Brief descriptions are given of preliminary repository design concepts and candidate site characteristics.

KEY WORD: Canada

C.46   Flow in Fractured Porous Media

        J. Duguid and P. Lei, ORNL and
        Princeton University, Princeton, N. J.

        Water Resources Research 13 (3),
        June 1977, p. 558

This report derives the equations governing
the flow of fluid through fractured porous
media: Darcy's Law for fluid flow in the
primary pores, equations of motion for fluid
flow in the fractures, and two continuity
equations. The transient flux of fluid out
of the primary pores and into the fractures
is described.

KEY WORD: transport

C.47   National Waste Terminal Storage
        Program Conference on Waste-Rock
        Interactions, Pennsylvania State
        University, July 1977

        Y/OWI/SUB-77-14268

This conference was generated in connection
with the NWTS Program. The technical reports include an overview of waste-rock
interaction studies and papers on stress-,
strain-, and temperature-induced permeability changes, the heater experiment at Oak
Ridge, shale consolidation and brine migration studies, and methods for determining
thermal history.

KEY WORDS: rock, geological properties

C.48   Calculations of Nuclide Migration
        in Rock and Porous Media, Penetrated
        by Water

        H. Haeggblom, Aktiebolaget
        Atomenergi, Stockholm, Sweden

        AE-RF-79-3264, September 1977

Physical and mathematical models are described
for migration of nuclides in rock and porous
media penetrated by water. Cases considered
are thermal convection due to the decay heat
from radioactive sources and transport due
to the hydraulic gradient associated with
geographic structure. Calculations, results,
and error margins are discussed.

KEY WORDS: transport, computer modeling

C.49   Nuclear Energy's Dilemma: Disposing
        of Hazardous Radioactive Waste
        Safely

        GAO

        EMD-77-41, September 1977

In this report to the Congress, GAO discusses the progress and problems of ERDA
in managing radioactive waste and the
problems of the NRC in handling accumu-
lating spent nuclear fuel. Included are
recommendations for regulatory and program
management changes.

KEY WORDS: waste management programs

C.50   The Nuclear Regulatory Commission
        Low-Level Radioactive Waste Man-
        agement Program

        NRC

        NUREG-0240, September 1977

This is a brief description of the NRC
program for management of low-level radioactive wastes which are disposed of in
commercial burial grounds. Preliminary
schedules and critical relationships
of the elements in the program are given.
Discussion includes 1) general program
and policy development, 2) standards
development for shallow-land burial and
alternate disposal methods, 3) regulations
and regulatory guides, and 4) development
of analytical models and staff review
procedures.

KEY WORDS: low-level waste, waste
            management programs

C.51   Storage and Release of Radiation
        Energy in Salt in Radioactive
        Waste Repositories

        G. Jenks and C. Bopp, ORNL

        ORNL-5058, October 1977

This study describes investigations of
amounts of gamma-ray energy stored in salt
under a variety of exposure conditions.
Also reported are thermal annealing

characteristics of the stored energy in salt, the chemical reactions undergone by radiation defects in salt upon aqueous dissolution of the salt, and the retention of the radiation defects within salt crystals.

KEY WORD: salt

C.52   Geological Criteria for Radioactive Waste Repositories

        G. Brunton and M. McClain, OWI

        Y/OWI/TM-47, November 28, 1977

This report gives a comprehensive list of geological criteria and factors for deriving specifications for evaluation of geological formations as repositories. Mechanical properties, state of stress, seismicity and site development are addressed, as well as socioeconomic criteria.

C.53   History of Prototype High-Level Waste Canister SS-9 While in Air and Water Storage

        D. Bradley, BPNL

        PNL-2278, November 1977

The history of a leak in prototype stainless steel canister filled with a high-level phosphate ceramic waste material containing 2 million curies of radioactivity is described. Details of each sampling period are provided.

C.54   High-Level and Long-Lived Radio-active Waste Disposal

        E. Angino, University of Kansas, Lawrence, KS

        Science 198 (4320), December 2, 1977, p. 885

This article discusses methods of radioactive waste burial in geologic formations, ice sheets, and seabed. Argument is made for further investigation of ice sheets and the ice-free areas of Antarctica and of arid zones as candidate waste sites.

C.55   Migration Paths for Oklo Reactor Products and Applications to the Problem of Geological Storage of Nuclear Wastes

        G. Cowan, LANL

        IAEA-TC-119/26; Proceedings of an IAEA Symposium on Natural Fission Reactors, Paris, France, December 1977

This paper describes a theoretical model for the escape of reactor products from the Oklo deposit. Assumptions are made regarding the modes of escape, loss rates via diffusion, and mobility. Results are applied to the problem of radioactive waste disposal.

KEY WORD: transport

C.56   Site Selection and Evaluation Studies of the Waste Isolation Pilot Plant (WIPP), Los Medaños, Eddy County, New Mexico

        G. Griswold, SLA

        SAND 77-0946, December 1977

This report describes site evaluation studies of the feasibility of constructing an underground facility for the terminal isolation of radioactive waste in south-eastern New Mexico salt beds. The scope of the report is limited to geologic investigation of the immediate area proposed for the site; as of June 1977. Many illustrations are included: surface and geologic maps, drilling maps, and photographs of the area and of site exploration activities.

KEY WORDS: bedded salt, geologic properties

C.57   Steady Thermal Convection at Low Rayleigh Number from Concentrated Sources in Porous Media

        C. Hickox, Jr., SLA

        SAND 77-1529, December 1977

This report describes several mathematical models for the analysis of steady thermal convection from a concentrated heat source in a fluid-saturated porous medium, as part of an assessment of the sea bed as an effective barrier to release of disposal nuclear waste. Results are presented for several idealized situations: point source, vertical line source, and isothermal sphere, imbedded in an unbounded porous medium.

KEY WORDS: thermal behavior

C.58 The Migration of Radioactive Substances Transported by the Groundwater from a Repository in a Rock

B. Grundfelt, KBS

December 1977

This report, one in a series by the Swedish Committee for Nuclear Safety, studies the migration of radionuclides from a repository for vitrified high-level waste in Swedish bedrock. The model developed at Battelle-Pacific Laboratories, comprises migration with flowing groundwater dispersion and geochemical retardation of nuclides. Descriptions of the model and system, physical parameters and model assumptions are given.

KEY WORDS: Sweden, rock, transport, computer modeling

C.59 Gold and Selective Storage Solves Nuclear Waste Problem

G. Wrangler, Royal Institute of Technology, Stockholm, Sweden

Annals of Nuclear Energy 4, 1977, p. 527

This article reports on a method of isolating transuranic elements in nuclear waste by using a thick layer of gold plate on the disposal canisters.

C.60 Site Selection Factors for Reposi-
*   tories of Solid High-Level and
      Alpha-Bearing Wastes in Geologic
      Formations

IAEA

ISBN 92-0-125177-7, 1977

This report offers general guidance for evaluating potential locations for repositories for radioactive waste, with emphasis on solid waste forms emplaced in deep geological formations. The selection factors discussed are topography, structure, physical and chemical properties, hydrology, future geological events, engineering conditions, and economic and social considerations.

KEY WORDS: waste management programs, high-level waste

C.61 Handling of Spent Nuclear Fuel
*   and Final Storage of Vitrified
      High-Level Reprocessing Waste
      (5 Vols.)

KBS

1977

The five volumes of this report describe a proposed method for the handling and storing of high-level waste, from fuel pools to final disposal in Swedish bedrock. The volumes include 1) investigations of suitability and test results for three geological sites, 2) descriptions of transportation systems and storage facilities, 3) an extensive safety analysis, and 4) discussions of the energy plans and waste management studies of several countries.

KEY WORDS: Sweden, waste management programs, rock

C.62 System Design for Retrieval of High-Level Wastes at Hanford

H. Wallskog, Rockwell Hanford Operations, Richland, WA

1977

This report describes a conceptual waste retrieval system designed for demonstrating the capability to retrieve a projected 36,000,000 gallons of radioactive salt cake and sludge wastes from underground storage tanks at Hanford.

KEY WORD: retrievability

C.63   Preliminary Assessment of the Radio-
        logical Protection Aspects of Dispo-
        sal of High-Level Waste in Geologic
        Formations

        M. Hill and P. Grimwood, NRPB

        NRPB-R69, January 1978

This is a preliminary study to assess the
potential radiological consequences of disposing of vitrified high-level radioactive
waste in geologic formations. A mathematical model of radionuclide migration in
groundwater is used to predict release
rates into fresh water from a hypothetical
repository containing all the high-level
waste which may be generated in the United
Kingdom up to the year 2000. The report
identifies areas of further study needed
for a full evaluation of this option.

KEY WORDS: radiological consequences,
              England

C.64   Status of Nuclear Fuel Reprocess-
        ing, Spent Fuel Storage, and
        High-Level Waste Disposal

        E. Varanini, III and R. Maullin,
        Nuclear Fuel Cycle Committee of
        the California Energy and Resources
        Conservation and Development Com-
        mission

        Draft, January 1978

This draft report on the status of nuclear
fuel reprocessing, spent fuel storage, and
high-level waste disposal is based on an
inquiry conducted by the Nuclear Fuel Cycle
Committee. It presents an analysis of the
status of technologies and issues in the
major portions of the back end of the nuclear fuel cycle. The investigation was
motivated by California legislature bills
charging the Energy Commission to determine
whether technologies for reprocessing and
waste storage exist and are acceptable for
approval by the appropriate federal agencies.

C.65   Modelling Studies Used to Evaluate
        Waste Disposal Options

        P. Grimwood, M. Hill, and G. Webb,
        National Radiological Protection
        Board, England

Nuclear Engineering International
23, January 1978, p. 55

Modeling studies are described which evaluate disposal of vitrified high-level waste
from reprocessing of nuclear fuels in geologic media and the deep ocean. Assumptions
and results are summarized.

C.66   Summary of Geologic Review Group
  *   Meeting, New Orleans, November 1977

        J. Frye, Geologic Society of
        America, Inc.

        Y/OWI/TM-18/4, January 1978

The Geologic Review Group was formed to study
and review 1) the long-term geologic stability of rock units considered for nuclear
waste disposal and 2) all geologic study
plans and activities of the NWTS Program
leading to site selection and development
of facilities for safe disposal of nuclear
waste. This report is a summary of their
findings and recommendations.

KEY WORDS: rock, waste management
              programs, salt

C.67   The Management of Canada's
  *   Nuclear Wastes

        Canadian Department of Energy,
        Mines, and Resources, Canada

        Atom 257, March 1978, p. 74

This article presents a synopsis of a study
of the Canadian nuclear program. The study
includes evaluations and recommendations
based on the once-through fuel cycle concept typified by the CANDU reactor, on
commercial reprocessing, on disposal, and
on interim waste management techniques.

KEY WORDS: Canada, waste management
              programs

C.68   Microstructural Interactions of
        Geologic Media with Waste
        Radionuclides

        T. Hinkebein and P. Hlava, SLA

        SAND-78-0108, March 1978

A preliminary investigation of the micro-structure of four geologic media (Magenta dolomite, Bell Canyon silt stone, Eleana shale, and clay-bearing halite) is described in terms of the interaction with Cs, Sr, Gd, and U. Elemental distribution photomicrographs of test ions and individual minerals are included.

KEY WORDS: geologic properties

C.69 Proceedings of a Symposium on
* Waste Management, Tucson, Arizona, March 1978

M. Wacks and R. Post, eds., University of Arizona, Tucson

University of Arizona, 1978

The papers in this collection address "Waste Management and Fuel Cycles". Sessions are on political, social, and public interactions; low-level waste criteria, controls, and management; volume reduction techniques; environmental concerns; and risk assessment.

C.70 Swedish-American Cooperative Pro-
* gram on Radioactive Waste Storage in Mined Caverns in Crystalline Rock - Program Summary

P. Witherspoon, LBL and O. Degerman, SKBF

LBL-7049/SAC-01, May 1978

This is the program summary for a series of reports documenting the results of the Swedish-American cooperative research program in which participating scientists explore the geological, geophysical, hydrological, geochemical, and structural effects associated with using a large crystalline rock mass as a repository for nuclear waste. The twenty-six technical reports in this series are abstracted and listed within the appropriate category. Page references are given in the title index under each title and also under the collective title of this entry.

KEY WORDS: granite, Sweden

C.71 Interactions Between Nuclear Waste and
* Surrounding Rock

G. McCarthy, W. White, R. Roy, B. Scheetz, S. Komarneni, D. Smith, and D. Roy, Pennsylvania State University, University Park, PA

Nature 273, May 18, 1978, p. 216

Experimental evidence is presented for the formation of the minerals weeksite and pollucite, by hydrothermal alternation of vitrified radioactive waste and by reaction of calcine with adjacent basalts or shales, respectively. Tests and results are described briefly.

C.72 The State of Geological Knowledge
* Regarding Potential Transport of High-Level Radioactive Waste from Deep Continental Repositories

EPA

EPA/520/4-78-004, June 1978

An ad hoc panel of earth scientists was charged with performing an evaluation of the adequacy of the state of knowledge in the earth sciences for reliably estimating the environmental impacts to be expected from the disposal of radioactive waste in deep geologic formations. This report was prepared to present the panel's findings to the EPA to provide guidance regarding the uncertainties inherent in estimates based on the state of knowledge.

KEY WORDS: transport, environment, radiological consequences

C.73 Nuclear Waste Canister Thermally-Induced Motion

P. Dawson and J. Tillerson, SLA

SAND 78-0566, June 1978

This is a detailed report on the development of a thermodynamically-coupled finite element model of viscous flow and heat transfer in salt due to heat-generating radioactive waste canisters. Analyses include factors such as temperature-dependent thermal conductivity, creeping viscous flow, and thermoelasticity.

KEY WORDS: salt, computer modeling

C.74   Development and Application of a
        Risk Assessment Method for Radio-
        active Waste Management. Volume 2.
        Implementation for Terminal Storage
        in Reference Repository and Other
        Applications

        S. Logan and M. Berbano, EPA,
        Las Vegas, NV

        EPA 520/6-78-005, July 1978

A systems model for radioactive waste management is presented in detail. In this
volume the first part of the systems model
consisting of the source term (radioactive
inventory vs. time), the release model, and
the environmental model, are described. The
candidate site for application is described
and application of the model to a repository
in shale is discussed. The environmental
model includes the transport and accumulation at various receptors in the biosphere,
pathways from these concentrations, and
resulting radiation dose to man. All elements of the system are given detailed
treatment.

KEY WORDS: risk analysis, radiological
              consequences, environment,
              computer modeling, transport

C.75   An Appraisal of Underground Radio-
        active Waste Disposal in Argillaceous
        and Crystalline Rocks: Some Geochemi-
        cal, Geomechanical, and Hydrogeologi-
        cal Questions

        J. Apps, N. Cook, and P. Witherspoon,
        LBL

        LBL-7047, Proceedings of a Symposium
        on Geotechnical Assessment and In-
        strumentation Needs in Crystalline
        and Argillaceous Rocks for Radioactive
        Waste Storage, Berkeley, CA, July 1978

This report is an appraisal of radioactive
waste disposal in argillaceous and crystalline
rocks, emphasizing mechanical, hydrologic, and
geochemical considerations. Factors affecting
site suitability are reviewed.

KEY WORDS: geochemistry, transport

C.76   Large Scale Permeability Test of
        the Granite in the Stripa Mine
        and Thermal Conductivity Test

        L. Lundstrom and H. Stille, KBS

The objective of the experiments described
here was to determine how the permeability
of an almost impermeable rock mass varies
with the pressure gradient, and the effective pressure and temperature at a special-
ly established test station with definable
boundary conditions. An investigation of
the thermal conductivity and effective
porosity of the rock mass was also carried
out on one of the four test places.

KEY WORDS: Sweden, granite, thermal
              behavior, permeability

C.77   Revised Concept for the Waste
 *   Isolation Pilot Plant

        A. Dennis, J. Milloy, L. Scully,
        H. Shefelbine, R. Stinebough
        and W. Wowak, SLA

        SAND 78-1429, July 1978

Facility design changes at WIPP resulting
from a reduction in the quantity of remote-
handled waste are described. Changes include those in the surface facility, and
receiving shaft, as well as the elimination of some buildings.

C.78   Radionuclide Interaction with Soil
        and Rock Media. Volume 1: Pro-
        cesses Influencing Radionuclide
        Mobility and Retention; Element
        Chemistry and Geochemistry; Con-
        clusions and Evaluation

        L. Ames and D. Rai, BPNL

        EPA 520/6-78-007, August 1978

This detailed report reviews the literature of interactions of 19 radionuclides
with soil and rock media to 1977. The
review includes a brief summary of natural

soil and rock distributions, chemistry, solid and solution equilibria and laboratory field adsorption and migration results. Suggestions for future work include determination of adsorption mechanisms and kinetics, influence of organic ligands on radionuclide migration potential, thermodynamic data and determinations of solid and solution species.

KEY WORDS: transport, geochemistry

C.79 An Analysis of the Measured Values for the State of Stress in the Earth's Crust

D. Jamison, LBL and N. Cook, UCB

LBL-7071/SAC-07, August 1978

Data from many experiments to measure the state of stress in the earth's crust have been analyzed. The information presented includes that from linear regression analyses and from interpretations of values of the coefficient of sliding friction between blocks of rock comprising the crust.

KEY WORDS: Sweden, granite

C.80 The Mechanical Properties of Stripa Granite

G. Swan, University of Lulea, Sweden

LBL-7074/SAC-03, August 1978

The mechanical properties of Stripa granite as determined from laboratory-size oven-dried specimens are presented. The properties determined include Poisson's ratio, uniaxial compressive fracture stress, and the coefficient of expansion, all as a function of temperature. The Brazilian tensile fracture stress, residual shear strength as a function of a normal stress, and the anisotropy ratios are also presented. Ultrasonic determinations of the rock's dilatational wave velocity at 1 MHz are given and the deduced Young's modulus is compared with the static value at room temperature.

KEY WORDS: Sweden, granite, mechanical behavior

C.81 Stress Measurements in the Stripa Granite

H. Carlsson, University of Lulea, Sweden

LBL-7078/SAC-04, August 1978

The Leeman three-dimensional overcoring method was used to determine the stress tensor for Stripa rock at the 348 m level. The experimental procedures and calculations are presented here.

KEY WORDS: Sweden, granite, mechanical behavior

C.82 Borehole Drilling and Related Activities at the Stripa Mine

P. Kurfurst, LBL; T. Persson, Hagby-Bruk AB, Nora, Sweden; G. Rudolph, VIAK AB, Falun, Sweden

LBL-7080/SAC-05, August 1978

This report describes the first stage of the Swedish-American Cooperative Program to investigate radioactive waste storage in mined caverns. Information is given on drilling operations, equipment, special techniques, layouts, scheduling, and costs.

KEY WORDS: Sweden, granite

C.83 Mining Methods Used in the Underground Tunnels and Test Rooms at Stripa

B. Andersson and P. Halen, Ställbergsbolagen, Ludvika, Sweden

LBL-7081/SAC-08, August 1978

This report describes smooth blasting and slot drilling, two methods tested and used in excavations at Stripa. A study of the fracturing caused by blasting is included.

KEY WORDS: Sweden, granite

C.84   A Pilot Heater Test in the
        Stripa Granite

        H. Carlsson, University of
        Lulea, Sweden

        LBL-7086/SAC-06, August 1978

Stress and temperature changes were moni-
tored at different radial distances from
a heater placed at the bottom of a 10 m
deep borehole. Measurements of movements
along major fractures on the surface and
changes of water in-flow to boreholes were
also carried out. This report gives a full
technical description of the pilot heater
test and results.

KEY WORDS: Sweden, granite, thermo-
              mechanical behavior

C.85   Geological Criteria for Reposi-
  *   tories for High-Level Radioactive
        Wastes

        Committee on Radioactive Waste
        Management, National Research Council

        National Academy of Sciences,
        August 1978

This brief report, prepared by a panel on
geological site criteria, presents a summary of the general criteria needed to
determine the suitability of geologic sites
for the storage or disposal of high-level
radioactive wastes. The study, requested
by the NRC, considers the use of an under-
ground-engineered structure in a suitable
geological environment. Qualitative, general criteria for site suitability are
presented.

KEY WORDS: geological properties

C.86   Criticality Analysis of Aggregations
        of Actinides from Commercial Nuclear
        Waste in Geological Storage

        E. Allen, ORNL

        ORNL/TM-6458, August 1978

This report provides information on the critical masses of the various isotopes present
in spent fuel or HLW. A computational model

was developed and critical mass calculations made for five waste types, five
waste ages, five actinide elements, and
four geologic compositions. The results
of the study were intended for use with
geologic transport rates to estimate mass
formation probabilities in waste repositories. The study was done in cooperation
with the NWTS program.

KEY WORDS: waste management programs,
              computer modeling, critical
              mass, actinide

C.87   Theoretical Temperature Fields for
        the Stripa Heater Project

        T. Chan, LBL; N. Cook, UCB;
        C. Tsang, UCB

        LBL-7082/SAC-09, September 1978

This report describes thermal conduction
calculations for the three in-situ heater
experiments at Stripa. A semi-analytic
solution using Green's function and a
numerical model using the integrated finite
difference technique have been applied to
the field situations at the Stripa site.
Detailed results are presented in tables,
temperature profiles and contour plots.

KEY WORDS: Sweden, thermomechanical
              behavior, computer modeling

C.88   Proceedings of a Seminar on In-
        Situ Heating Experiments in Geo-
        logical Formations, Ludvika/Stripa,
        Sweden, September 1978

        OECD and SKBF

        OECD, 1978

The objectives of the seminar were to
review existing information on in-situ
heating experiments in geological formations, to exchange ideas, and to promote
cooperation. The state of the art is represented in the papers, covering clay,
salt, and crystalline rocks, and providing
details of the experiments at Stripa.

KEY WORD: Sweden

C.89   Management of Radioactive Fuel
        Wastes: The Canadian Disposal
        Program

        J. Boulton, ed.; Atomic Energy
        of Canada Research Company,
        Pinawa, Manitoba, Canada

        AECL-6314, October 1978

This report describes the research and development program to verify the concepts
for the safe, permanent disposal of radioactive wastes from Canadian nuclear reactors.
Discussion includes the nature of the waste,
and options for storing, processing, packaging, and disposing of the waste. The program to verify the proposed concept, to
select a suitable site, and to build and op-
erate a demonstration facility is described.

KEY WORDS: Canada, waste management
              programs

C.90   Mechanical and Thermal Design
        Considerations for Radioactive
        Waste Repositories in Hard Rock

        N. Cook, LBL, and P. Witherspoon, UCB

        LBL-7073/SAC-10, October 1978

The report contains two papers which discuss
the fundamental considerations of thermomechanical aspects of repository design.
Part I, "An Appraisal of Hard Rock for Potential Underground Repositories of Radioactive Waste", analyzes data on the mechanical safety and stability of underground repositories. Part II, "In Situ Heating Experiments in Hard Rock: Their Objectives
and Design" describes an experiment for
obtaining information on the effects of heating a large volume of rock.

KEY WORDS: Sweden, granite, thermal
              behavior, mechanical behavior

C.91   Risk Methodology for Geologic Dis-
        posal of Radioactive Waste: Sensi-
        tivity Analysis Techniques

        R. Inman, J. Helton, and
        J. Campbell, SLA

        NUREG/CR-0394, SAND 78-0912
        October 1978

Statistical techniques for sensitivity
analysis of a complex model, including
hypercube sampling, partial rank corre-
lation, rank regression and predicted
error sum of squares, are presented. The
application is the analysis of a model
for the surface movement of radionuclides.

C.92   Risk Methodology for Geologic
        Disposal of Radioactive Waste:
        Interim Report

        J. Campbell, R. Dillon, M. Tierney,
        H. Davis, and P. McGrath, SLA;
        F. Pearson and H. Shaw, USGS;
        J. Helton and F. Donath, Univer-
        sity of Illinois

        NUREG/CR-0458, SAND 78-0029,
        October 1978

This report documents the first phase of
the development of a methodology for
assessing the long-term risks from radioactive waste disposal in deep geologic
media. Models and limited databases are
given, as well as detailed discussion of
the hypothetical reference system used
to provide a setting for exercising the
models.

KEY WORDS: risk analysis

C.93   Decommissioning of Surface Facili-
        ties Associated with Repositories
        for the Deep Geological Disposal
        of High-Level Nuclear Wastes

        R. Heckman, LLNL

        IAEA-SM-234, November 1978

A methodology is presented to evaluate decommissioning of the surface facilities
associated with repositories for the deep
geologic disposal of high-level nuclear
waste. A cost/risk index is described
and proposed as an evaluation criterion.
The modes considered are protective storage, entombment, and dismantlement. An
example using a preliminary baseline repository design is given.

KEY WORDS: decommissioning

C.94   International Nuclear Fuel
 *   Cycle Evaluation

         IAEA

         Reports of the First Plenary Con-
         ference of the International Nuclear
         Fuel Cycle Evaluation (INFCE), Vienna,
         November 1978, 1980

The eight working groups of this conference
studied topics including reprocessing,
spent fuel management, and waste management
and disposal. The study of waste management
and disposal analyzes wastes from seven fuel
cycles: LWR (once-through and U-Pu cycle),
FBR (U-Pu cycle), HWR (once-through, U-Pu
cycle, and U-Th cycle), and HTR (U-Th cycle).
Options for waste repositories for all waste
arisings are considered. The reference repositories are deep geological formations,
salt or granite. Environmental impacts are
discussed.

KEY WORDS: salt, granite

C.95   Fracture Detection in Crystalline
         Rock Using Ultrasonic Shear Wave

         K. Waters and S. Palmer, UCB, and
         W. Farrell, Systems, Science, and
         Software, La Jolla, CA

         LBL-7051/SAC-19, December 1978

This report describes the development of a
seismic profiling system which uses re-
flected shear waves to detect artificial
and natural cracks in large laboratory rock
specimens. Results of laboratory experiments are presented.

KEY WORDS: Sweden, granite, seismic
              safety

C.96   Full-Scale and Time-Scale Heating
         Experiments at Stripa: Preliminary
         Results

         N. Cook, LBL and M. Hood, UCB

         LBL-7072/SAC-11, December 1978

Two full-scale heating experiments for
assessing the near-field effects of thermal
loading for the design of an underground
nuclear waste repository are described. A

time-scale heating experiment to obtain
field data of the interaction between
heaters and its effect on the rock mass
during a two-year period is also described.

KEY WORDS: Sweden, granite, thermal
              behavior

C.97   Geochemical Constraints on Accumu-
         lation of Actinide Critical Masses
         from Stored Nuclear Waste in
         Natural Rock Repositories

         D. Brookins, ONWI

         ONWI-17, December 1978

This is a collection of reports on lanthanide and actinide individual and joint sy-
stematics. The papers selected cover such
topics as uranium solution chemistry,
uranium deposits, the Oklo, West Africa,
reactor, rare-earth deposits, manganese
nodules, and bedded and domal salt deposits.

KEY WORDS: rock, salt, geochemistry,
              actinides

C.98   Methodology for the Determination
         of Environmental Iodine 129 and
         Technetium 99

         T. Anderson, SRL

         DP-MS-77-75X, 1978

This is a summary of the work at SRL to
develop analytical methods to determine
the environmental impact of the SRL plant
with respect to Iodine 129 and Technetium
99. Items of concern are 1) concentration in soil and vegetation as a function
of distance from the plant, 2) vertical
migration from the soil surface, and 3)
underground lateral transport from radioactive waste basins and storage sites.

KEY WORD: environment

C.99   Safety Assessment of Radioactive
         Waste Disposal into Geological
         Formations. A Preliminary Appli-
         cation of Fault-Tree Analysis to
         Salt Deposits

         B. Bertozzi, M. D'Alessandro,
         F. Girardi, M. Vanossi, CEC

EUR 5901 EN, 1978

This study is an application of fault-tree analysis to geological repositories of radioactive wastes. Criteria and approaches to the quantification of probabilities of primary events are addressed. The assessment of probability values for release events due to natural causes and human actions up to one million years is discussed. Two ideal cases of saline formations are used for illustrative purposes in the analysis.

KEY WORDS: radiological consequences

C.100 Report to the President
  *
       IRG

       TID-28817 (Draft), 1978

This report presents the findings, policy considerations, and tentative recommendations of the IRG as a result of its consideration of alternative strategies identified for nuclear waste management. The report, issued for public comment, includes technical, institutional, and management issues.

KEY WORDS: waste management programs, regulations, risk analysis

C.101 Alternative Technology Strategies
  *   for the Isolation of Nuclear Waste -
       Subgroup Report

       IRG

       TID-28818 (Draft), 1978

This report provides information on alternative technological strategies that could lead to a disposal facility for radioactive wastes. The strategies involve the following candidate disposal options: 1) emplacement in mined repositories, 2) emplacement in deep ocean sediments, 3) emplacement in very deep drill holes, 4) emplacement of reprocessing waste in a mined cavity leading to rock melting, 5) partitioning of reprocessing waste, transmutation of heavy radionuclides and geologic disposal of fission products, and 6) ejection into space. The strategies are presented and analyzed from selected standpoints, including health

and safety consequences, economic considerations, and assurances of technical and programmatic successes.

KEY WORDS: waste management programs, radiological consequences, risk analysis, high-level waste, transuranium waste

C.102 Geological Disposal of High-Level Radioactive Wastes - Earth-Science Perspectives

       J. Bredehoeft, A. England, D. Stewart, N. Trask, and I. Winograd, USGS

       USGS Circular 779, 1978

This report is an examination of the earth science problems associated with geologic disposal of high-level radioactive wastes. Issues discussed are repository host rocks and site characterization, perturbations resulting from waste emplacement, the groundwater system, containment time frames and geologic prediction. Further research needs are cited.

KEY WORDS: geologic properties, rock

C.103 Technical Support for GEIS: Radioactive Waste Isolation in Geologic Formations, 23 Volumes

       Science Applications, Inc., Oak Ridge, TN; Dames & Moore, N. Y. and Parsons, Brinckerhoff, Quade, & Douglas, N. Y.

       Y/OWI/TM-36, 1978

This series of documents were prepared under the direction of OWI to supplement Y/OWI/TM-44, "Contribution to Draft Generic Environmental Impact Statement on Commercial Waste Management: Radioactive Waste Isolation in Geologic Formations." The series provides further technical support for the preconceptual designs, resource requirements, and environmental source terms associated with isolating commercial LWR wastes in underground repositories in salt, granite, shale, and basalt. Wastes from three fuel cycles are considered: uranium and plutonium recycling, no spent fuel recycling, and uranium-only recycling.

KEY WORDS: waste management programs,
              bedded salt, granite, shale,
              basalt

C.104   Handling and Final Storage of Unre-
        processed Spent Nuclear Fuel (2 Vols.)

        KBS

        1978

This report, by a group formed by the Swedish
power utilities, describes the safe final
storage of spent unreprocessed nuclear fuel.
The first, general volume gives premises
and data, a description of the steps of
handling procedures, a summary of dispersal
processes, and a safety analysis. The
second, technical volume contains detailed
sections on geology and the facilities in
the handling chain. The safety analyses
include discussions of transportation, health
effects, and acts of war and sabotage, as
well as dispersal calculations.

KEY WORDS: Sweden, waste management
              programs

C.105   Safe Disposal of High-Level
  *   Nuclear Reactor Wastes: A
        New Strategy

        A. Ringwood, Books Australia,
        Norwalk, CT

        1978

This book presents a method for producing
synthetic igneous rock systems, SYNROC, as
a host material for the safe disposal of
high-level nuclear reactor wastes. Experimental basis is given for claiming that
radioactive waste is better confined in
synthetic rocks than in natural crystalline
igneous and metamorphic rocks.

KEY WORDS: high-level waste, SYNROC,
              geochemistry, rock

C.106   The Geological Criteria for Suit-
  *   able Sites of High-Level Radioactive
        Waste Repositories

        Committee on Radioactive Waste Man-
        agement National Research Council

        National Academy of Sciences, 1978

This study by a committee of the National
Research Council Commission on Natural Re-
sources considers the use of a carefully
engineered structure located deep underground in a suitable geological environment.
Qualitative discussions of geological and
geochemical properties of a potential repository are given. The study includes a
preliminary set of general criteria for
repository site suitability.

KEY WORDS: geological properties

C.107   Radionuclide Migration in Aerated
        Zones. 3. Theoretical Estima-
        tion of Strontium 90 Migration

        S. Morisawa, Y. Inoue, and
        Y. Mahara; Kyoto University, Kyoto,
        Japan, and Central Research Insti-
        tute of Electric Power Industry

        Nihon Genshiryoku Gakkaishi 20,
        1978, p. 133

This report describes a method for estimating radionuclide migration through an
aerated zone with consideration of unsteady
soil moisture infiltration process. The
method was examined against experimental
results on Sr 90 distribution in a model
aerated sand bed. Results are given in
detail and conclusions tabulated.

KEY WORDS: transport

C.108   Electrical Heaters for Thermo-
        mechanical Tests at the Stripa
        Mine

        R. Burleigh, E. Binnall, A.
        DuBois, D. Norgren, and
        A. Oritz, LBL

        LBL-7063/SAC-13, January 1979

This report describes three types of
heaters designed, fabricated, and installed at Stripa. Details of the systems are given, as well as discussions
of experience gained during testing, in-
stallation, and operation.

KEY WORDS: Sweden, granite,
              instrumentation

C.109   The Status of Borehole Plugging and
         Shaft Sealing for Geologic Isola-
         tion of Radioactive Waste

         D'Appolonia Consulting
         Engineers, Inc.

         ONWI-15, January 1979

This report reviews and evaluates penetra-
tion sealing (borehole plugging and shaft
sealing) activities and research which have
been conducted in the U. S. and Europe.
Important technical considerations for
repository sealing are identified.

C.110   Monitoring Instrumentation Spent
         Fuel Management Program

         EA&T

         UCRL 15060, January 1979

Preliminary monitoring system methodologies
are identified as an input to the risk
assessment of spent fuel management. Conceptual approaches to instrumentation for
surveillance of canister position and orien-
tation, vault deformation, spent fuel dissolution, temperature, and health physics conditions are presented.

KEY WORDS: monitoring, instrumentation

C.111   Limitations to the Use of Two-
         Dimensional Thermal Modeling of
         a Nuclear Waste Repository

         B. Davis, LLNL

         UCID 18101, January 1979

This report examines recent modeling studies
which show that the time-dependent temperature distribution can be accurately modeled
in the far-field using a 2-D planar numerical model. The near-field cannot be modeled
accurately enough by either 2-D axisymmetric
or 2-D planar numerical models for repositories in salt. The accuracy limits of 2-D
modeling were defined by comparing results
from 3-D TRUMP modeling with results from
both 2-D axisymmetric and 2-D planar. Both
TRUMP and ADINAT were employed as modeling
tools. Two-dimensional results from the
finite element code, ADINAT, were compared
with 2-D results from the finite difference
code, TRUMP.

KEY WORDS: computer modeling, thermal
             behavior

C.112   Fluid Inclusions in Salt - An
  *   Annotated Bibliography

         D. Isherwood, LLNL

         UCID 18102, January 1979

This report is an annotated bibliography
of information on fluid inclusions in salt.
Limited data from the "Salt-Vault" in situ
heater experiments in the early 1960's are
noted to give little doubt that fluid inclusions can migrate towards a heat source.
There is also a brief summary of the physical and chemical characteristics that,
together with the temperature of the waste,
will determine the chemical composition of
the brine in contact with the waste canister,
the rate of fluid migration, and the brine-
canister-waste interactions.

KEY WORDS: inclusions, bedded salt,
             brine migration

C.113   Convection and Thermal Radiation
         Analytical Models Applicable to a
         Nuclear Waste Repository Room

         B. Davis, LLNL

         UCID 18103, January 1979

This report 1) identifies the thermodynamic
properties and physical parameters of three
convection regimes -- forced, natural, and
mixed; 2) defines the convection corela-
tions applicable to calculating heat flow
in a ventilated (forced-air) and in a non-
ventilated nuclear waste repository room;
and 3) describes a computer code that com-
putes and compares the floor-to-ceiling
heat flow by convection and radiation and
determines the nonlinear equivalent conductivity tables for a repository room.

KEY WORDS: thermal behavior, computer
             modeling

C.114   Earth Science Technical Plan for
  *   Mined Geological Disposal of
         Radioactive Waste

         DOE and USGS

         TID-29018, Draft, January 1979

This technical plan examines earth science questions related to nuclear waste management. The report identifies the components of a comprehensive program necessary to resolve the questions and to make progress towards a licensed repository. The report includes a detailed and systematic review of earth science issues, an inventory and classification of current R&D to facilitate integration and coordination of DOE and USGS-sponsored work, and recommends future R&D directions.

KEY WORDS: geological properties

C.115   Environmental and Other Evaluations
*   of Alternatives for Long-Term Manage-
        ment of Stored INEL Transuranic
        Waste

        DOE

        DOE/ET-0081, February 1979

This document discusses all aspects of waste management options for INEL, including a detailed description of the Radioactive Waste Management Complex established in 1952 as a controlled area of about 144 acres for burial of solid radioactive waste generated by INEL operations.

KEY WORD: environment

C.116   High-Level Waste Repository Site
        Suitability Study - Status Report

        R. Heckman, D. Towse, D. Isherwood,
        T. Harvey, and T. Holdsworth, LLNL

        NUREG/CR-0578, UCRL 52633,
        February 1979

This report describes work on site suitability criteria for SHLW repositories. It includes 1) a physical model which simulates the natural environment of many potential sites, 2) mathematical models for the calculation of the performance of hypothetical repository sites, 3) a parameter database representative of the natural environment, and 4) analyses and results.

KEY WORDS: computer modeling,
             database

C.117   Data Acquisition, Handling, and
        Display for the Heater Experiment
        at Sripa

        M. McEvoy, LBL

        LBL-7062/SAC-14, February 1979

This report describes the data acquisition system used to acquire and digitize data, convert measured values to engineering units, store data, provide on-site graphical displays and transfer data for further analysis of information gathered through the heater experiments at Stripa. Details are given of the system, its design, capabilities, and operation.

KEY WORDS: Sweden, instrumentation

C.118   Rock Mass Characterization for
        Storage of Nuclear Waste in Granite

        P. Witherspoon, P. Nelson, T. Doe,
        R. Thorpe, and B. Paulsson, LBL,
        and J. Gale and C. Forster, Uni-
        versity of Waterloo, Ontario,
        Canada

        LBL-8570/SAC-18, February 1979

This report describes rock mass characterization at Stripa by four different methods: 1) mechanical characterization, including monitoring jointed rock response to thermal loading, 2) geological characterization including surface, subsurface, and core mapping, 3) geophysical characterization using several borehole techniques, and 4) hydrologic characterization, through injection tests, pump tests, water pressure measurements, and controlled inflow tests.

KEY WORDS: Sweden, granite, mechanical
             behavior, hydrologic proper-
             ties, geologic properties

C.119   National Waste Terminal Storage
        Program - ONWI Technical Program
        Plan

        ONWI

        ONWI-19, February 1979

This report presents a summary of the over-all program for the development and operation of a geologic nuclear waste repository. An overview is given of objectives, technical scope, milestones, budget, and organizational and management processes.

KEY WORDS: waste management programs

C.120   Evaluation of Cement Borehole Plug
        Longevity

        D. Roy, M. Grutzeck, P. Licastro,
        Pensylvania State University,
        University Park, PA

        ONWI-30, February 1979

This report describes considerations for assessing the longevity of cement-based materials used for waste repository bore-hole plugging and shaft sealing. The dominant factors addressed are: the extent of attainment of equilibrium and rate of approach of the plug component chemical subsystem to a state of stable equilibrium with the total plug-rock-groundwater system; the effect upon accompanying changes in physical, mechanical, and thermal properties of the plug-rock system; and its consequent effectiveness in preventing radionuclide transport.

KEY WORD: transport

C.121   Proceedings of a Symposium on
*   Waste Management, Tucson,
        Arizona, February-March 1979

        M. Wacks and R. Post, eds.,
        University of Arizona, Tucson

        University of Arizona, 1979

Papers at this symposium address the state of waste disposal technology and its social and political implications. Sessions are on the technology of treatment, criteria for evaluating waste management systems, and the special problems of nuclear waste.

KEY WORDS: waste management programs

C.122   Geologic Migration Potentials of
        Technetium 99 and Neptunium 237

        E. Bondietti and C. Francis, ORNL

Science, 203, March 30, 1979,
p. 1337

This report reviews risk assessments of technetium and neptunium as potentially capable of migrating from high-level radio-active waste repositories. Experimental results are given in detail.

KEY WORD: transport

C.123   Properties of Radioactive Calcine
        Retrieved from the Second Calcined
        Solids Storage Facility at Idaho
        Chemical Processing Plant

        B. Staples, G. Porniak, and
        E. Wade, Idaho Chemical Processing
        Plant, Idaho

        ICP-1189, March 1979

This report describes measurements of the chemical and physical properties of radio-active alumina and zirconia calcine samples retrieved from the storage bins at the Idaho Chemical Processing Plant. Data were compared with properties of samples taken during calcination of nonradioactive calcine. Results of the testing program are presented.

C.124   Seismic Safety in Nuclear
        Waste Disposal

        D. Carpenter and D. Towse, LLNL

        UCID 18125, April 26, 1979

This report reviews the data on damage to underground equipment and structures from earthquakes, the record of associated motions, and the conventional methods of seismic safety-analysis and engineering. Safety considerations are divided into two classes: those during the operational life of disposal facility, and those pertinent to the post-decommissioning life of the facility. The report describes techniques which need to be developed to address the question of long-term earthquake probability in relatively aseismic regions, and for discriminating between active and extinct faults in regions where earthquake activity does not result in surface ruptures.

KEY WORDS: seismic safety

C.125   Environmental Impact Statement:
  *   Waste Isolation Pilot Plant,
         2 Volumes

         DOE

         DOE/EIS-0026-D, April 1979

This document was prepared as environmental
input for decisions on the development and
implementation of a safe and acceptable
radioactive waste management program. The
focus of this statement is WIPP, and decisions concerning a transuranic waste repository, an intermediate-scale facility,
and associated experiments. These two
volumes contain objectives and alternatives
for meeting the objectives, interim waste-
acceptance criteria, a detailed description
and environmental analysis of the WIPP repository and proposed site, and a summary
section.

KEY WORDS: waste management programs,
              transuranic waste,
              environment

C.126   Environmental Impact Statement:
  *   Management of Commercially Genera-
         ted Radioactive Waste, 2 Volumes

         DOE

         DOE/EIS-0046-D, April 1979

This environmental impact statement replaces
WASH-1539 (C.14) on the program for developing interim and permanent repositories for
high-level and transuranic radioactive
wastes. This statement considers comments
on WASH-1539 as well as comments from other
public sectors and technical conferences.
These two volumes include technology alternatives for final disposal of waste and com-
parative analyses of options.

KEY WORDS: waste management programs,
              transuranic waste, environment

C.127   Geochemistry and Isotope Hydrology
         of Groundwaters in the Stripa
         Granite: Results and Preliminary
         Interpretation

         P. Fritz, J. Barker, and J. Gale,
         University of Waterloo, Ontario,
         Canada

         LBL-8285/SAC-12, April 1979

Groundwater samples collected from the
Stripa site, from shallow, private wells,
surface boreholes, and boreholes drilled
from the 330 m and 410 m mine levels were
analyzed for their major ion chemistry,
dissolved gases, and isotope content. The
report presents a summary and preliminary
interpretation of the chemical and isotope
data.

KEY WORDS: Sweden, granite,
              geochemistry

C.128   Three-Dimensional Thermal Analysis
  *   of a High-Level Waste Repository

         T. Altenbach, LLNL

         UCID 17984, April 1979

This report documents the three-dimensional
thermal analysis of a high-level waste repository. The analysis used the TRUMP computer code to evaluate the thermal fields
for six repository scenarios that studied
the effects of room ventilation, room
backfill, and repository thermal diffusivity. The results for selected nodes are
presented as plots showing the effect of
temperature as a function of time.

KEY WORDS: computer modeling, thermal
              behavior

C.129   Retrieval System for Emplaced
         Spent Unreprocessed Fuel (SURF)
         in Salt Bed Depository Baseline
         Concept Criteria Specifications
         and Mechanical Failure Probabilities

         IECO

         UCRL 15025, May 1979

This report describes the entire retrieval
scenario of degraded SURF canisters from
subterranean depository to surface and
presents criteria specifications on each
piece of retrieval equipment.

KEY WORDS: retrievability, bedded
              salt

C.130   An Approach to the Fracture
         Hydrology at Stripa: Prelimi-
         nary Results

         J. Gale, University of Waterloo,
         Ontario, Canada, and P. Witherspoon,
         LBL

         LBL-7079/SAC-15, May 1979

The approach used to integrate the information gathered from an extensive series
of boreholes and fracture maps at Stripa
is reviewed and preliminary results from
the fracture hydrology are given.

KEY WORD: Sweden

C.131   Preliminary Report on Geophysical
         and Mechanical Borehole Measurements
         at Stripa

         P. Nelson, B. Paulsson, R. Rachiele,
         L. Andersson, T. Schrauf, LBL, and
         W. Hustrulid, TerraTek, Salt Lake
         City, Utah

         LBL-8280/SAC-16, May 1979

Extensive detail is provided on four sets of
borehole measurements made at the Stripa
site during 1978, using four different measurement systems: 1) a continuous logging
unit equipped with natural gamma, neutron,
gamma-gamma, temperature, electrical resistivity, caliper, and sonic probes, 2) mechanical systems for the measurement of
mechanical moduli, 3) ultrasonic equipment
for detecting fractures, and 4) borehole logging equipment emphasizing electrical techniques for detecting fractures.

KEY WORDS: Sweden, granite, measurement,
              instrumentation

C.132   Observations of a Potential Size-
         Effect in Experimental Determina-
         tion of the Hydraulic Properties
         of Fractures

         P. Witherspoon, C. Amick, and
         J. Gale, LBL, and K. Iwai, UCB

         LBL-8571/SAC-17, May 1979

Preliminary results are given for studies
on the effect of stress on fracture permeability properties for rock specimens
of various sizes. Discussion is focused
on the effect of sample size on these
results.

KEY WORDS: Sweden, granite

C.133   Technical Concept for Test of
         Geologic Storage of Spent Reactor
         Fuel in the Climax Granite, Nevada
         Test Site

         L. Ramspott, L. Ballou, R. Carlson,
         D. Montan, T. Butkovich, J. Duncan,
         W. Patrick, D. Wilder, W. Brough,
         and M. Mayr, LLNL

         UCID-18197, May 1979

The Spent Fuel Test in the Climax granite
at the Nevada Test Site is a generic test
in which spent fuel assemblies from an
operating commercial nuclear reactor are
emplaced at and retrieved from a plausible
waste repository depth in a typical granite.
This report describes the site, the preop-
erational measurement program, field instrumentation, data acquisition systems,
and the techniques for handling the spent
fuel. Thermal and mechanical response calculations are summarized.

KEY WORD   granite, thermal behavior,
               mechanical behavior,
               measurement

C.134   U. S. Program for the Immobilization
         of High-Level Nuclear Wastes

         J. Crandall, SRL

         DP-MS-79-2X/ CONF-790602; American
         Nuclear Society Spring Meeting,
         Atlanta, GA, June 1979

This paper proposes a program for immobilization of waste. The elements described
are 1) development and choice of management
alternatives, 2) waste form and characterization.

KEY WORDS: waste management programs

C.135   Gas Generation from Radiolytic
         Attack of Transuranic-Contaminated
         Hydrogenous Waste

         A. Zerwekh, LANL

         LA-7674-MS, June 1979

An investigation is reported of the possibility that alpha-radiolysis of hydrogenous
materials might produce toxic, corrosive,
and flammable gases in retrievable-stored
waste. Several levels of contamination were
studied, as well as pressure, temperature,
and moisture effects. Experiments are described and results given.

KEY WORDS: transuranic waste

C.136   Past Climate Reconstruction: A Tool
         for Assessing Site Suitability

         G. Potter, LLNL

         UCID 18118, June 1979

This report uses the method of reconstruc-
turing past climate variations to arrive
at a better understanding of possible future
precipitation and groundwater recharge patterns. This work provides input into hydrologic modeling efforts for waste management
programs. Both short-term (0 to 350 y) and
long-term (∿ 6000 y ago) estimates are given.

KEY WORDS: climate, hydrologic properties

C.137   Proceedings of a Workshop on Thermo-
         mechanical Modeling for a Hard Rock
         Waste Repository, June 25-27, 1979,
         Berkeley, CA

         F. Holzer and L. Ramspott, eds.,
         LLNL

         ONWI-98, 1979

The objectives of this workshop, sponsored
by ONWI and arranged by LLL, were: 1) to
identify the key issues associated with
modeling and validating the response of a
hard rock repository to emplaced waste and
2) to identify activities which would serve
as programmatic input to ONWI and DOE. The
proceedings summarize discussions on modeling and calculations, laboratory and field
measurements, instruments and measurement

techniques, and field experimentation for
model validation.

KEY WORDS: hard rock, thermomechanical
              behavior, computer modeling,
              waste management programs

C.138   Alternatives for Long-Term Man-
  *   agement of Defense Transuranic
         Waste at the Savannah River Plant,
         Aiken, South Carolina

         DOE

         DOE/SR-WM-79-1, July 1979

Risks and costs are presented for 12 alternatives for the long-term management of defense transuranic waste now stored in
trenches and on concrete pads at the Savannah River Plant. Analyses include concerns
of occupational exposure and non-radiologi-
cal fatal injuries to workers during construction and retrieval.

KEY WORDS: risk analysis

C.139   Characterization of Discontinuities
         in the Stripa Granite Time-Scale
         Heater Experiment

         R. Thorpe, LBL

         LBL-7083/SAC-20, July 1979

The methodology and results of a detailed
study of geologic discontinuities associated with the time-scale heater experiment at Stripa are given. The object of
the report is to define the position and
the nature of major discontinuities in the
heated region. Characteristics of the
fractures, such as types and thicknesses
of infilling, sizes, and spacings, are
discussed statistically. Some hypotheses
regarding the origin of the fracture system
are suggested.

KEY WORDS: Sweden, granite

C.140   Some Results from a Field Investi-
         gation of Thermomechanical Loading
         of a Rock Mass When Heaters Are
         Emplaced in the Rock (I) and the
         Application of Field Data from
         Heater Experiments Conducted at

Stripa. Sweden, to Parameters for
Repository Design (II)

M. Hood, UCB, and H. Carlsson and
P. Nelson, LBL

LBL-9392/SAC-26, July 1979

In Part I results are presented of a field
experiment to monitor the response of a rock
mass to thermomechanical loading from elec-
trically heated canisters emplaced in the
rock at a depth of 340 m. Measurements are
made of temperature, displacement, and stress
fields. Part II focuses on two specific considerations affecting repository design:
limits on canister power levels in the near
field as imposed by decrepitation of the
borehole wall and the ability to predict the
thermally-induced stresses and their impact
on far-field effects. Both problems are discussed in terms of field results from the
experimental program at Stripa.

KEY WORDS: Sweden, granite, thermo-
             mechanical behavior

C.141   Characterization of Samples of a
        Cement-Borehole Plug in Bedded
        Evaporites

        B. Scheetz, M. Grutzeck, L. Wakeley,
        and D. Roy, The Pennsylvania State
        University, University Park, PA

        ONWI-70, July 1979

This report describes the laboratory characterization of a section of a cement borehole
plug in the Salado Formation near Carlsbad,
New Mexico. Characterization of the plug and
host rock was carried out by a combination of
measurements including compressive strength,
permeability, density and porosity, thermal
measurements, x-ray measurements, and optical
microscopy.

KEY WORDS: bedded salt, measurement

C.142   Effects of Temperature, Temperature
        Gradients, Stress, and Irradiation
        on Migration of Brine Inclusions in
        a Salt Repository

        G. Jenks, ORNL

        ORNL-5526, July 1979

This is a review and analysis of available
experimental and theoretical information
on brine migration in bedded salt, including the effects of temperature, thermal
gradients, stress, irradiation, and pressure. Detailed comparisons and analyses
are presented.

KEY WORDS: thermal behavior, brine
             migration, inclusions

C.143   Proceedings of the NEA Workshop
        on the Use of Argillaceous Materials
        for the Isolation of Radioactive
        Waste, Paris, September 1979

        OECD

        OECD, 1979

Session papers summarize research and development activities relevant to the disposal of radioactive wastes in argillaceous
formations. Topics include parameters
affecting radionuclide migration, in-situ
geochemical properties of clays subject to
thermal loading and mathematical models
for clay behavior in nuclear waste storage
problems.

KEY WORD: migration

C.144   A Model for the Transport of
        Radionuclides and Their Decay
        Products Through Geologic Media

        H. Burkholder and E. Rosinger,
        Battelle, Columbus and Atomic
        Energy of Canada, Ltd.

        ONWI-11, September 1979

The transport of radionuclides and their
decay products from an underground nuclear
waste repository through the surrounding
media to a surface environment is modeled
in one dimension. The mathematical solution is applicable to the evaluation of
the sensitivity of potential releases of
radioactivity to the characteristics of
various nuclear waste isolation systems.

KEY WORDS: computer modeling,
             transport

C.145   Computer Modeling of Nuclear Waste
 *   Storage Canister Corrosion

        D. Cottrell, W. Ludemann, and
        D. McCright, LLNL

        UCRL-83488, October 18, 1979

This report describes a way to better predict
the corrosion process with computer modeling.
A program is developed to calculate anodic
and cathodic polarization curves using Tafel
slopes, equilibrium exchange current densi-
ties, and other electro-chemical parameters
obtained from the experimental corrosion
literature. The model generates and displays
polarization curves for different values of
environmental parameters such as temperature,
pH, and concentrations of pertinent species
in the vicinity of the canister material.

KEY WORDS: computer modeling,
               corrosion

C.146   Considerations for Development of
        Specifications for Subsurface Waste
        Handling Equipment

        J. Arbital, E. Bettis, T. Myrick,
        H. Watts, R. Wilems, H. Yook,
        Science Applications, Inc.,
        Oak Ridge, TN

        ONWI-75, October 1979

This report presents the results of a systems
engineering approach to developing a methodology for generating criteria and specifications for subsurface waste handling equipment.
Key decisions and analyses required to complete development of detailed criteria and
specifications are discussed qualitatively.

C.147   Thermal Gradient-Brine Inclusion
        Migration in Salt Study Gas-Liquid
        Inclusions - Preliminary Model

        D. Olander and A. Machiels, UCB

        ONWI-85, October 1979

A model is presented for the behavior of two-
phase inclusions in a thermal gradient in a
salt medium. The migration analysis includes
the calculation of temperature distributions,
water flux, and inclusion velocity.

KEY WORDS: brine migration,
               inclusions, thermal
               behavior

C.148   Retrieval System for Emplaced
        Spent Unreprocessed Fuel (SURF)
        in Salt Bed Depository

        IECO

        UCRL-15111, October 1979

This report provides support in developing
an accident prediction event tree diagram,
with an analysis of the baseline design
concept for the retrieval of emplaced spent
unreprocessed fuel (SURF) contained in a
degraded canister. The report contains
an evaluation check list, accident logic
diagrams, accident event tables, fault
trees/event trees and discussions of failure probabilities for the following sub-
systems as potential contributors to a
failure: 1) canister extraction, including
the "core" and "ram" units, 2) canister
transfer at the hoist area, and 3) canister hoisting.

KEY WORDS: retrievability, bedded
               salt

C.149   Storage of Spent Nuclear Fuel,
        Papers from ASCE National
        Convention, Atlanta, GA,
        October 1979

This collection of selected papers includes
discussions of environmental, geotechnical,
and structural issues associated with the
interim and long-term storage of spent
nuclear feel.

C.150   Waste Isolation Performance Assess-
        ment and In-Situ Testing - Pro-
        ceedings of the U.S./FRG Bilateral
        Workshop, Berlin, October, 1979

        ONWI

        ONWI-88, 1979

This report contains reprints of papers
from the U.S./German waste repositories
and related in-situ experiments. The
papers cover topics such as performance

assessment, WIPP, brine migration experiments, and mine atmosphere studies.

KEY WORDS: Germany, waste management
programs

C.151 Proceedings of the National Waste Terminal Storage Program Information Meeting, October-November, 1979, Columbus, Ohio

DOE

ONWI-62, 1979

This report is a collection of papers presented at an NWTS Information Meeting organized as one of a series of meetings to provide a review of the progress of ONWI programs, for members of the technical community and the general public. Sessions are on technical studies in the NWTS/ONWI Science and Technology Program, geologic studies in the NWTS/ONWI Site Identification Program, and technical studies in programs on process/equipment development, systems analysis, site and repository licensing, and facilities engineering.

KEY WORDS: waste management programs

C.152 Alternative Waste Disposal Concepts-An Interim Technical Assessment

D. Crandall, Bechtel National, Inc., San Francisco, CA

ONWI-65, November 1979

This report is the final product of the Alternative Waste Disposal Concepts Study, part of DOE's NWTS program. The study identifies and describes five alternative waste disposal concepts. Cases developed from combinations of these alternatives with two different spent fuel forms are assessed with respect to 1) radiological impact, 2) degree of development, 3) resource consumption, 4) safeguards, and 5) economics. A tentative ranking of the specific cases and recommendations for further research are provided.

KEY WORDS: waste management programs,
environment, safeguards

C.153 Status Report on the Importance of Natural Organic Compounds in Groundwater as Radionuclide-Mobilizing Agents

J. Means and D. Hastings, Battelle, Columbus

ONWI-84, December 1979

Preliminary studies on radionuclide mobilization by organic complexing agents are described. Results are presented for chemical analyses of uranium solution mine leachate and deep groundwater from Hanford. The report includes literature reviews on 1) the origen and geochemistry of humic substances, 2) the aqueous transport of actinides by humic substances, and 3) the enrichment of actinides by organic substrates.

KEY WORDS: transport, geochemistry

C.154 Review of Current Capabilities for the Measurement of Stress, Displacement, and In Situ Deformation Modulus

T. Schrauf and H. Pratt, TerraTek, Salt Lake City, Utah

ONWI-95, December 1979

Current capabilities for the measurement of stress, displacement, and the in situ deformation modulus are reviewed with respect to accuracy, sensitivity, advantages, and limitations. Both instruments and measurement techniques are considered. Recommendations are made for adapting existing measurement techniques to repository monitoring. These include the development and modification of gages, extensometers, and stress meters.

KEY WORDS: measurement, instrumentation

C.155 Review of Geotechnical Measurement Techniques for a Nuclear Waste Repository in Bedded Salt

IECO

UCRL-15141, December 1979

This report presents a description of geotechnical measurement techniques that can provide the data necessary for safe development --i.e., location, design, construction, operation, decommissioning and abandonment-- of a radioactive waste repository in bedded salt.

The techniques discussed in this report are grouped in the following categories: 1) geologic, geophysical, and geodetic, 2) rock mechanics, 3) hydrologic, hydrogeologic, and water quality, and 4) thermal.

The report presents extensive tables that provide a review of available measurement techniques for each of these categories. The tables give the purpose of the measurements, the applicable repository development stage, a brief description of the techniques, and references in which detailed discussions can be found.

KEY WORDS: measurement, bedded salt

C.156   Geoscience Parameter Database Handbook: Granites and Basalts

Willard Owens Associates, Inc., Wheat Ridge, CO

UCRL-15152, December 1979

This report documents the known geologic parameters of large granite and basalt occurrences in the coterminous United States, for future evaluation in the selection and licensing of radioactive waste repositories. The description of the characteristics of certain potential igneous hosts has been limited to existing data pertaining to the general geologic character, geomechanics, and hydrology of identified occurrences.

KEY WORDS: basalt, rock, granite, geologic properties

C.157   Proceedings of a Symposium on Decommissioning of Nuclear Facilities, Vienna, November 1978

IAEA and OECD

IAEA, 1979

Sessions in this symposium are on policy and standard development for decommissioning nuclear facilities; engineering considerations; radiological release factors and waste classification, decommissioning experience, including industrial experience gained in decontamination and partial dismantling of a shut-down reprocessing plant; decontamination and remote operations.

KEY WORDS: decommissioning, radiological consequences

C.158   Analyses of the Effect of Variations in Parameter Values on the Predicted Radiological Consequences of Geologic Disposal of High-Level Waste

M. Hill, NRPB

NRPB-R86, 1979

This preliminary assessment of radiological consequences of geologic disposal of high-level waste deals with the effect of groundwater transport of radionuclides. Parameters considered include the leach rate of the waste, groundwater velocity, length of flow path, and sorption constants of the principal radionuclides. Results of the sensitivity analysis are used to identify research priorities, to establish preliminary repository design criteria, and to indicate developments needed in mathematical modeling of radionuclide transport.

KEY WORDS: transport, radiological consequences, computer modeling

C.159   Migration of Long-Lived Radionuclides in the Geosphere

OECD

Proceedings of the Workshop Organized by OECD/NEA and CEC, Brussels, Belgium, 1979

These papers present the technical aspects of research programs in several countries to investigate the migration of radionuclides from radioactive wastes buried in geological formations. Topics covered are transport models, migration, retention, plutonium, and the environment.

KEY WORDS: transport, migration, geologic
              properties, computer modeling
              radiological consequences

C.160   The Hydrologic Properties of Shales

         California State University,
         Hayward, CA

         UCRL-15149, 1979

Shales and related rocks, siltstones, mud-stones, and claystones, belonging to a group of broadly represented impermeable rocks, are described as offering potential for long-term storage of radioactive wastes. This report provides an updated review of what is known about the hydrologic properties of shales, what can be presently guessed, and what remains to be done.

KEY WORDS: shales, hydrologic properties

C.161   NWTS Criteria for the Geologic
  *   Disposal of Nuclear Wastes: Site-
         Qualification Criteria

         ONWI

         ONWI-33(2), January 1980 (Draft)

This report describes general repository site-qualification criteria that DOE applies to the NWTS Program. The criteria address the issues of public health and safety, en-vironmental protection, engineering feasi-bility, institutional and socioeconomic im-pacts, and cost considerations. Project-specific information (e.g., information for Basalt Waste Isolation Project or Nevada Nuclear Waste Storage Investigations) will be developed consistent with these criteria and will be issued separately as appendixes to this report.

KEY WORDS: waste management programs

C.162   A Repository Post-Sealing Risk
         Analysis Using Macro

         A. Kaufmann, L. Edwards, and
         W. O'Connell, LLNL

         UCRL-82236, February 14, 1980

MACRO, a code to propagate probability dis-tributions through a set of linked models, is currently under development at Lawrence Livermore Laboratory. An early version of this code, MACRO1, has been used to assess post-sealing dose to man for simple repository and site models based on actual site data.

KEY WORDS: risk analysis, computer
              modeling

C.163   Progress with Field Investiga-
         tions at Stripa

         P. Witherspoon, LBL; N. Cook, UCB;
         and J. Gale, University of Waterloo,
         Ontario, Canada

         LBL-10559/SAC-27, February 1980

This report gives the results of hydro-radiological and thermomechanical experiments at Stripa. The problems investigated are 1) predicting the thermomechanical be-havior of a heterogeneous and discontinuous rock mass and 2) predicting the movement of groundwater that can transport radio-nuclides through the granite.

KEY WORDS: Sweden, granite, hydrologic
              properties, thermomechanical
              behavior, transport

C.164   Potential U.S./Canadian Cooperative
         Activities in Geological Disposal
         of Radioactive Waste

         J. Duguid, ONWI

         ONWI-19, March 1980

Meetings between staff from the U. S. Department of Energy and the Atomic Energy of Canada, Limited, were held in June, 1979, to discuss cooperative programs in waste isolation. Working groups discussed admin-istrative details, geotechnical topics such as field testing and exploration, and issues related to model development and assessment. The results of the delibera-tions are given in this report.

KEY WORD: Canada

C.165   Retrieval Options Study

        Kaiser Engineers, Inc., Oakland, CA

        ONWI-63, March 1980

This study is part of the systems analysis activities of ONWI to develop the scientific and technical bases for radioactive waste repositories in geologic media. A methodology and database are developed which allow the relative evaluation of retrieval and recovery costs. Technical criteria discussed are safety, feasibility, ease of retrieval, probable intact retrieval time, safeguards, monitoring, criticality, and licensability. This detailed report defines 505 repository options and evaluates costs and technical criteria comprehensively.

KEY WORDS: retrievability

C.166   Finite-Length Line Source Super-
        Position Model (FLLSSM)

        Kaiser Engineers, Inc., Oakland, CA

        ONWI-94, March 1980

This report documents a linearized thermal conduction model developed to determine temperatures of host media for geologic nuclear waste repositories. The methodology is outlined and the computer code which performs calculations is described in detail.

KEY WORDS: computer modeling, thermal
                behavior

C.167   Geologic Factors in the Isolation
        of Nuclear Waste: Evaluation of
        Long-Term Geomorphic Processes and
        Catastrophic Events

        S. Mara, SRI International,
        Menlo Park, CA

        PNWL-2854, March 1980

This report is part of the Pacific Northwest Laboratory program for the assessment of the effectiveness of geologic isolation systems. The information in this volume is input to a computer model used to simulate release scenarios and the consequences of release of nuclear waste from geologic containment. The long-term geomorphic processes evaluated are denudation, entrenchment, and aggradation. Catastrophic events were evaluated to determine their significance to the simulation model.

KEY WORDS: geologic properties,
                computer modeling

C.168   Proposed Rulemaking on the Storage
  *   and Disposal of Nuclear Waste -
        Statement of the Position of the
        U. S. Department of Energy

        DOE

        DOE/NE-0007, April 1980

This proceeding describes a nuclear waste management program which has specific performance objectives and a conservative approach to insure that the objectives will be met. Mined geologic disposal is selected as an interim planning strategy. The report assesses the degree of assurance now available that radioactive waste can be safely disposed of. Its purpose is to determine when such disposal or off-site storage will be available and to determine whether radioactive wastes can be safely stored on-site past the expiration of existing facility licenses until off-site disposal or storage is available. The various possible methods for storage or disposal of spent fuel are assessed, and the basis for selecting a preferred method is described. Each component of the preferred system is discussed individually. The factors important to its function are described, each proposed or existing requirement for its performance is stated, and the current ability of the program to meet such requirements is discussed. Where there are technial uncertainties concerning the ability to meet requirements, the report describes the research and development still to be carried out and how these results are expected to resolve the uncertainties.

KEY WORDS: waste management programs

C.169   Environmental Impact Statement -
  *   U. S. Spent Fuel Policy, Final,
        5 Volumes

        DOE

        DOE/EIS-0015, May 1980

The five volumes of this EIS include detailed discussions of 1) storage of U. S. spent power reactor fuel, 2) storage of foreign spent power reactor fuel, 3) charges due for storage of spent fuel accepted from foreign sources, 4) comments on draft EIS, and 5) DOE responses to major comments. Several appendixes in each volume contribute to the vast quantity of information in this report on U. S. policies, facilities, and environmental impacts.

KEY WORDS: environment, waste management programs

C.170 Repository Sealing: Evaluations of Materials Research Objectives and Requirements

D'Appolonia Consulting Engineers, Pittsburgh, PA

ONWI-108, June 1980

This is a detailed proposal for a program for materials research in support of the NWTS program for repository sealing. Priorities are assigned to focus the effort. Engineering properties of a seal considered included chemical, hydraulic, mechanical, and thermal properties and emplacement criteria. Several candidate sealing materials are reviewed.

C.171 Logistics Characterization for Regional Spent Fuel Repositories Concept

D. Joy, ORNL, and B. Hudson, ONWI

ONWI-124, August 1980

This report summarizes logistics considerations for a four-region waste repository system: northeastern, north central, southern, and western regions of the U. S. The logistics considerations include yearly receipt and emplacement, inventory, AFR storage, nuclear capacity growth effects, reactor lifetimes, proportions of PWR/BWR fuel, proportions of rail and truck shipments, shipping cask fleet requirements, number of annual shipments, and shipment costs.

C.172 Preliminary Investigation of the Thermal and Structural Influence of Crushed-Salt Backfill on Repository Disposal Rooms

R. Wagner, RE/SPEC, Inc., Rapid City, SD

ONWI-138, August 1980

This report describes investigations of the thermal and structural influence of crushed salt backfill within a typical room in a radioactive waste repository. The study is part of the NWTS program rock mechanics investigations.

C.173 Environmental Development Plan-
* Defense Waste Management

DOE

DOE/EDP-0064, September 1980

This environmental development plan is the basis for planning and reviewing the progress of DOE's program for defense waste management. The report identifies the planning and management requirements and schedules needed to evaluate and assess the environmental, health, and safety aspects of the defense waste management program. The technology program and environmental strategy are described in detail, with further information on regulations and research provided in appendixes.

KEY WORDS: waste management programs, military waste, environment

C.174 Retrievability: Technical Considerations

R. Wilems, J. Arbital, J. Fowler, J. Mosier, L. Rikertsen, and P. Stevens, Science Applications, Inc., Oak Ridge, TN

ONWI-203, September 1980

This study furthers work on a methodology developed to relate the length of a retrievability period to technical considerations. Key technical considerations are identified and analyzed in detail from the perspective of site selection, design and construction of a waste repository, waste emplacement, retrievability, decommissioning, and post-decommissioning.

KEY WORD: retrievability

C.175   Technical Conservatisms in NWTS Repository Conceptual Designs: National Waste Terminal Storage Repository No. 1 - Special Study No. 4

         Stearns-Roger Services, Inc., Denver, CO

         ONWI-222, September 1980

This report identifies and explains the major technical conservatisms in the conceptual designs for NWTS program repositories. Areas discussed include thermal loading of the geologic structure, rock mechanics and underground design, waste throughput capacity, hoisting systems, nuclear criticality safety, confinement of radioactive materials, occupational exposure and health physics, environmental effects and cost estimates.

C.176   Comparison of Potential Radiological Consequences from a Spent-Fuel Repository and Natural Uranium Deposits

         O. Wick and M. Cloninger, BPNL

         PNL-3540, September 1980

This report investigates the rationale of the criterion that deep geologic repositories containing spent fuel should impose no greater radiological risk than that due to naturally occurring uranium deposits. An analysis is carried out to determine whether current expectations of spent fuel repository performance are consistent with that criterion. In this report, comparisons are based on 1) intrinsic characteristics, such as radionuclide inventory, depth, proximity to acquifers, and regional distribution, and 2) actual and potential radiological consequences now occurring from some ore deposits and that may eventually occur from repositories.

KEY WORDS: radiological consequences

C.177   Review of Potential Host Rocks for Radioactive Waste Disposal

         SRL

         DP-1559, DP-1561, DP-1562, DP-1563, DP-1564, DP-1567, DP-1568, DP-1569, October 1980

As part of the NWTS Program, studies related to the storage of waste in the geologic environment of southeastern United States, i.e., in the igneous and metamorphic rocks of the Piedmont, the sands and clays of the Coastal Plains, and the mudstones and shales of the Triassic basins from Maryland to Georgia. Literature reviews were done to designate areas potentially suitable for disposal of solidified high-level radioactive waste. The series of detailed reports from which SRL prepared a summary report (DP-1559) is as follows:

Piedmont Province of Virginia and Maryland; W. Brown, University of Kentucky, DP-1561.

Piedmont Province of North Carolina; J. Butler, University of North Carolina, DP-1562.

Piedmont Province of South Carolina, D. Secor, Jr., University of South Carolina, DP-1563.

Piedmont Province of Georgia; D. Wenner and K. Gillon, University of Georgia, DP-1564.

Southern Piedmont Subregion; Acres American, Inc., Buffalo, N. Y., DP-1567.

Southeastern Coastal Plain Subregion; Ebasco Services, Inc., New York, N. Y., DP-1568.

Triassic Basin Subregion; Dames and Moore, Inc., Atlanta, Georgia, DP-1569.

KEY WORDS: rock, hydrologic properties, geologic properties

C.178   Desicant Materials Screening for Backfill in a Salt Repository

         D. Simpson, Lehigh University, Bethlehem, PA

ONWI-214, October 1980

This report describes an experimental study of MgO and CaO to determine their suitability as desicants for backfilling around a nuclear waste repository. Experimental procedures and results are presented in detail.

C.179   Summary Characterization and Recommendation of Study Areas for the Gulf Interior Region

Bechtel National Inc., San Francisco, CA and Law Engineering Testing Co., Marietta, GA

ONWI-18, November 1980

This report further characterizes the Gulf Interior Region with respect to environmental and geological information. See also "Regional Environmental Characterization Report for the Gulf Interior Region and Surrounding Territory" (C.180).

C.180   Regional Environmental Characterization Report for the Gulf Interior Region and Surrounding Territory

Bechtel National Inc., San Francisco, CA

ONWI-67, November 1980

This report is part of the regional study phase of the NWTS program. It contains detailed maps and environmental information for the Gulf Interior Region. The data in the report meet the requirements of predetermined survey plans and will be used to identify areas that will be further characterized. Information on the physical environment, demography, socioeconomics, land use, and the biological environment is presented.

KEY WORDS: environment

C.181   Regional Environmental Characterization Report for the Paradox Bedded Salt Region and Surrounding Territory

Bechtel National, Inc. San Francisco, CA

ONWI-68, November 1980

This environmental characterization of the Paradox Basin salt region is a product of the NWTS Program. The report includes general information from published sources on the physical environment, socioeconomics, demography, the biological environment, and regulatory requirements. The data will be used in determining smaller areas for further characterization.

KEY WORDS: environment, bedded salt, hydrologic properties, geologic properties

C.182   Risk Assessment Methodology Development for Waste Isolation in Geological Media

C. Stevens, R. Fullwood, and S. Basin, SAI

NUREG/CR-1672, SAI-219-80-PA, November 1980

This is an independent review of three volumes of work, NUREG/CR-0458, -0394, and -0424, by Sandia Laboratories on the development of a methodology for assessing the long-term risk of a nuclear waste repository in a geologic medium. This report, by SAI, provides an overview of the documents, a review and critique of each, and recommendations for further testing and development.

KEY WORDS: risk analysis

C.183   Management of Radioactive Waste Gases from the Nuclear Fuel Cycle - Volume I: Comparison of Alternatives

A. Evans, W. Prout, J. Buckner, M. Buckner, SRL

NUREG/CR-1546, December 1980

This study reports on alternatives for the collection and fixation of radioactive waste gases released during normal operation of the nuclear fuel cycle and for the transportation, storage, and disposal of the resulting waste forms. Integrated waste management schemes are compared for krypton-85, iodine-129, and carbon-14. A projected Volume II will deal with performance criteria for the waste forms associated with the waste gases.

KEY WORDS: waste management programs

C.184   Estimated Environmental Effects
         of Deep Drilling

         R. McPherson, D. Waite, D. Shipler
         and M. Glora, ONWI

         ONWI-13, December 1980

This report describes impacts on the environment associated with deep drilling
activities performed to collect data on
geologic formations considered candidate
waste repository sites by the NWTS Program.
Impacts varying in type, permanence, and
intensity are outlined.

KEY WORD: environment

C.185   Estimated Environmental Effects
         of Geologic and Geophysical
         Exploratory Activities

         R. McPherson, D. Waite, D. Shipler,
         and M. Glora, ONWI

         ONWI-105, December 1980

This report summarizes effects on the environment due to NWTS Program activities, such
as shallow drilling, geohydrologic testing,
well logging, seismic testing, electrical
resistivity surveys, gravity surveys, mag-
netic surveys, geologic mapping, and micro-
earthquake network studies.

KEY WORD: environment

C.186   Technical Summary and Index for
         National Waste Storage Repository
         No. 1 and No. 2: Terminal Concep-
         tual Design Reports, Special Study
         No. 6

         Stearns-Roger Services, Inc.,
         Denver, CO

         ONWI-234, December 1980

This is part of the NWTS Program study of
repository conceptual design. This report
contains an overview of the conceptual design reports, glossary, and technical index,
all focussed on facilitating use of the
multi-volume conceptual design reports.

C.187   Uptake and Retention of Radionu-
         clides by Marine Organisms: A
         Bibliography Prepared for the
         Seabed Program

         V. Schultz, Washington State
         University, Pullman, WA

         SAND 80-7133, December 1980

This is a bibliography on marine radiation
ecology prepared for the Seabed Disposal
Biology Program at Sandia Laboratories,
New Mexico. References are given alpha-
betically, by author, with no further
organization.

KEY WORDS: marine environment

C.188   Assay of Brines for Common
         Radiolysis Products

         C. MacDougall, ORNL

         ORNL/TM-7568, January 1981

This report describes experimental work on
the products formed by radiolysis of concentrated solutions of brines. Micro-
analytic techniques for free Cl_2, H_2O_2,
ClO_3^-, and ClO_4^- were developed, adapted,
and tested. Explicit directions for assays
are detailed and preparation of reagents
and standards described in appendixes.

C.189   NWTS Program Criteria for Mined
         Geologic Disposal of Nuclear
         Waste: Site Performance Criteria

         DOE

         DOE/NWTS-33(2), February 1981

This document is part of the NWTS-33
series, providing guidance for the NWTS
Program. This volume contains criteria
governing the suitability of sites for
mined geologic disposal of high-level
radioactive waste. Summarized are criteria
associated with site geometry, geohydrolo-
gy, geochemistry, geologic characteristics,
tectonic environment, human intrusion, surface characteristics, demography, environmental protection and socioeconomic impacts.

KEY WORDS: waste management
              programs

C.190   KBS Annual Report - 1980 Including
         Summary of Technical Reports
         Issued During 1980

         KBS

         March 1981

This report provides a summary of the technical activities of KBS, now a permanent
part of SKBF. Studies in many areas of
the long-term behavior and safety of repositories for radioactive wastes are reported
on: materials, chemistry, engineered barriers, geology, hydrology, modeling of nuclide migration, low- and medium-level
wastes, and the Stripa project, and foreign
information exchange agreements.

KEY WORDS: waste management programs

C.191   International Conference on Nuclear
         Waste Transmutation, July 22-24,
         1980, Austin, Texas

         University of Texas
         Austin, Texas

         March 1981

These proceedings include conference papers
from sessions on transmutation of radwaste,
risk assessment and fuel cycle impacts,
partitioning and separation of wastes, nuclear cross sections and computational
methods. Included also are the keynote
address and panel discussions on all topics,
reproduced in dialogue format.

KEY WORDS: risk analysis, transmutation

C.192   Sensitivity and Uncertainty Analysis
         of a Simple, Near-Field Granite
         Nuclear Waste Repository

         Y. Ronen, J. Lucious, E. Oblow,
         ORNL

         ORNL/TM-7514, March 1981

An uncertainty analysis of nuclear waste
repository design parameters is described
using a sensitivity methodology based on
calculating by direct means the first and
second order sensitivity coefficients of
performance parameters with respect to input
data. The methodology was applied to a
simple near-field single level granite

repository model to determine uncertainties
in several significant repository temperatures. Input parameters are given in an
appendix.

KEY WORDS: computer modeling, granite

C.193   Calculated Thermally-Induced Dis-
         placements and Stresses for Heater
         Experiments at Stripa, Sweden-
         Linear Thermoelastic Models Using
         Constant Material Properties

         T. Chan, LBL, and N. Cook, UCB

         LBL-7061/SAC-22

This report describes the results of calculations of thermally-induced displace-
ments and stresses used to guide the
design, operation, and data interpretation
of in-situ heating experiments in a granite
formation at Stripa. The calculations are
done using linear thermoelastic finite
element models in which the rock mass is
assumed to be a homogenous, isotropic con-
tinuum with temperature-independent material
properties.

KEY WORDS: Sweden, rock, thermo-
               mechanical behavior,
               computer modeling

C.194   Alternative Processes for Managing
         Existing Commercial High-Level
         Radioactive Wastes

         BPNL

         NUREG-0043

This study identifies and reviews potentially viable options considered by ERDA
for the neutralized waste stored at ERDA
sites and examines the applicability of
these options to the management of neutralized wastes from facilities of Nuclear
Fuel Services, Inc. The report was designed to be useful to the NRC as a basis
for technical assessment of managing
existing wastes. Each waste management
option is examined with respect to its
technological status, research and development required, safety considerations, and
cost and time estimates.

KEY WORDS: waste management programs,
               high-level waste

C.195   Project Salt Vault: A Demonstration
        of the Disposal of High Activity
        Solidified Wastes in Underground
        Salt Mines

        ORNL

        ORNL-4555

This is a collection of technical papers on
the Project Salt Vault experiment. Included
are reports on studies preliminary to the
experiment and descriptions of laboratory
pillar model experiments.

KEY WORD: salt

C.196   Geoscience Database Handbook for
        Modeling a Nuclear Waste Reposi-
        tory (2 Volumes)

        D. Isherwood, LLNL

        UCRL 52719, NUREG/CR-0912

This handbook contains reference information on parameters that should be considered
in analyzing or modeling a proposed nuclear
waste repository site. Those parameters and
values that best represent the natural environment are included. Volume 1 contains a
database on salt as a repository medium.
Chapters on the geology of bedded and dome
salt, the geomechanics of salt, hydrology,
geochemistry, natural and manmade features
and seismology provide compiled data and
related information useful for studying a
proposed repository in salt. Volume 2 is
the result of a scoping study for a database
on the geology, geomechanics, and hydrology
of shale, granite, and basalt as alternative
repository media.

KEY WORDS: database, bedded salt, shale,
               basalt, granite, geologic
               properties, hydrologic proper-
               ties, seismology

C.197   Material Transport Through Porous
  *   Media: A Finite-Element Galerkin
        Model

        J. Duguid and M. Reeves, ORNL

        ORNL-4928

This report describes a two-dimensional
model of the flow of a dissolved consti-
tuent through porous media. Mechanisms
included in the model are advective
transport, hydrodynamic dispersion,
chemical adsorption, and radioactive
decay. The report outlines the development of the model in detail and gives
two examples.

KEY WORDS: transport, computer modeling

C.198   Long-Term Risk Assessment of
        Radioactive Waste Disposal
        in Geological Formations

        F. Girardi, G. Bertozzi, and
        M. D'Allessandro, CEC

This is a review of the work at the Joint
Research Center in Italy to study methods
for long-term safety analysis of nuclear
waste from power production in the European
Community. The methodology reviewed is
based on the assessment of the quantitative value of a system of barriers between
waste and man. The barriers considered
are the geological formation of the site,
the chemical and physical stability of the
waste form, subsoil retention, and the biosphere. Information is provided on environmental contamination, and pathways to
man.

KEY WORDS: risk analysis, radiological
            consequences, environment

C.199   Proceedings of the Symposium
        on Waste Management, Tucson,
        Arizona, October 1976

        (SEE ABSTRACT A.6.)

C.200   Management of Radioactive
        Wastes from the Nuclear
        Fuel Cycle, 2 Volumes

        IAEA

        Proceedings of a Symposium
        Organized by IAEA and OECD/NEA,
        Vienna, 1976, ISBN 92-0-020276-4

        (SEE ABSTRACT A.8.)

C.201   Alternatives for Managing Wastes
        from Reactors and Post-Fission
        Operations in the LWR Fuel Cycle,
        5 Volumes

        ERDA

        ERDA 76-43, 1976

        (SEE ABSTRACT A.9.)

C.202   Alternatives for Long-Term Manage-
        ment of Defense High-Level Radio-
        active Waste - Savannah River
        Plant, 2 Volumes

        ERDA

        ERDA 77-42, 1977

        (SEE ABSTRACT A.16.)

C.203   Alternatives for Long-Term
        Management of Defense High-
        Level Radioactive Waste -
        Idaho Chemical Processing Plant

        ERDA

        ERDA 77-43, 1977

        (SEE ABSTRACT A.17.)

C.204   Alternatives for Long-Term
        Management of Defense High-Level
        Radioactive Waste, Hanford
        Reservation

        ERDA

        ERDA 77-44, 1977

        (SEE ABSTRACT A.18.)

C.205   Determination of Performance Cri-
        teria for High-Level Solidified
        Nuclear Waste

        J. Cohen, LLNL

        NUREG 0279, 1977

        (SEE ABSTRACT A.20.)

C.206   Proceedings of the Symposium on
        "Science Underlying Radioactive
        Waste Managemenet," Materials
        Research Society, Boston, MA,
        1978. Scientific Basis for
        Nuclear Waste Management,
        Volumes 1 and 2

        G. McCarthy, ed., Vol. 1
        Materials Research Laboratory,
        Pennsylvania State University,
        University Park, PA

        C. Northrup, ed., Vol. 2, SLA

        (SEE ABSTRACT A.26.)

C.207   Assumptions and Ground Rules
        Used in Nuclear Waste Projections
        and Source Term Data

        S. Storch and B. Prince, Union
        Carbide Corporation, Oak Ridge, TN

        ONWI-24, September 1979

        (SEE ABSTRACT A.29.)

D. SOCIETAL, POLITICAL, AND ECONOMIC ISSUES OF WASTE MANAGEMENT

D.1 Proceedings of a Symposium on The
√ Management of Radioactive Wastes
From Fuel Reprocessing, Paris,
November-December 1972

OECD and IAEA

OECD, March 1973

These symposium papers deal with waste management problems associated with the reprocessing of irradiated fuel elements. An overview of the situation in this field is presented as well as a review of research underway. Papers include topics such as local releases of radionuclides and their regional and global impacts; practices and policies in foreign countries; techniques of solidification; storage and disposal methods.

KEY WORDS: radiological consequences, reprocessing

D.2 Environmental Survey of the
Uranium Fuel Cycle

AEC

WASH-1284, April 1974

This document is an update of the report "Environmental Survey of the Nuclear Fuel Cycle" published in 1972 by the Fuels and Materials Staff of the AEC Directorate of Licensing. It takes into account comments submitted in response to the Federal Register notice as well as recommendations for improvements made during rule-making hearings. The survey and hearings were limited to the uranium fuel cycle. Extensive

technical detail is provided on uranium mining and milling, uranium hexafluoride production, uranium enrichment, fuel fabrication, irradiated fuel reprocessing, waste management, and transportation.

KEY WORDS: environment, reprocessing

D.3 High-Level Radioactive Waste Management Alternatives - 4 Volumes

K. Schneider, A. Platt, eds.,
BPNL

BNWL-1900, May 1974

These nine sections issued as four volumes are an overview of potential alternative methods for long-term management of high-level radioactive waste. The sections are discussions on background and database evaluation methodology, geologic disposal, ice sheet disposal, sea bed disposal, waste partitioning, extraterrestial disposal, and transmutation processing. The study takes into account not only currently available technology but that which can be developed or is expected to be available in the near future.

KEY WORDS: waste management programs

D.4 Reactor Safety Study - An Assess-
* ment of Accident Risks in U. S.
Commercial Nuclear Power Plants

AEC

WASH-1400-D, August 1974

This report, also called the Rasmussen Study, assesses the risks to the public from potential accidents in nuclear power plants of the type being built in the United States. The study considers uncertainty in present knowledge and the resulting range in predictions. Specific objectives of the study include performing a realistic and quantitative risk assessment, developing risk methodologies and providing an independent check on the effectiveness of reactor safety practices in industry and government.

KEY WORDS: risk analysis

D.5   Isolating High-Level Radioactive Waste from the Environment: Achievements, Problems and Uncertainties

GAO

RED-75-309, December 1974

This is the Comptroller General's Report to the Congress reviewing and evaluating AEC's progress in assuring safe storage of high-level radioactive waste from all sources. The report contains no recommendations or suggestions, but is intended to be helpful to the Congress in its oversight of nuclear programs.

KEY WORDS: waste management programs

D.6   Radioactive Waste Management in Japan

R. Kiyose, University of Tokyo, Tokyo, Japan

CONF-751044; Nuclear Safety, 1975, Proceedings of an American Nuclear Society Topical Meeting, Tucson, Arizona, October 1975

This paper reviews the status of radioactive waste management in Japan. Problems addressed include treatment of wastes, policies on cement-solidification of low-level waste, research and development programs, environmental and social impacts, costs, and appropriate management agencies.

KEY WORDS: Japan, waste management programs

D.7   Radioactive Waste Management
  *   and Regulation

M. Willrich, MIT

MIT-EL 76-011, December 1976

This report is directed towards assisting in the development of public policy and institutions necessary for the safe management of high-level and TRU radioactive wastes. It reports on results of an MIT research project for ERDA and provides recommendations for institutional reforms to deal more effectively with radioactive wastes.

KEY WORDS: transuranic waste, waste management programs

D.8   A Review of Methods for the Detection of 10 nCi/g of Transuranic Isotopes in Solid Waste

W. King, LLNL

UCRL-52200, January 7, 1977

This report reviews the analyses and techniques available for low-level radiation detection. The problems of measuring transuranic isotopes in solid waste material at 10 nCi/g are discussed, including calculational considerations and methods of measuring higher allowable minimum levels in solid wastes.

D.9   Survey of Naturally Occurring Ha-
  *   zardous Materials in Deep Geologic Formations: A Perspective on the Relative Hazard of Deep Burial of Nuclear Wastes

K. Tonnessen and J. Cohen

UCRL-52199, January 14, 1977

Hazards associated with deep burial of solidified nuclear waste are considered in relation to the hazards of toxic elements in naturally occurring ore deposits. The basis for comparison derives from a consideration of safe drinking water levels. Calculations for relative toxicity of fast breeder reactor (FBR) waste and light water reactor (LWR) waste in an underground repository are compared with the relative toxicity indices obtained for average concentration ore deposits.

KEY WORDS: toxicity

D.10   Public Comments and Task Force Responses Regarding Environmental Survey of the Reprocessing and Waste Management Portions of the LWR Fuel Cycle (NUREG-0116)

NRC

NUREG-0216, Supplement 2 to WASH-1248, March 1977

This report is issued as part of the record in support of the proposed rule change to 10CFR Part 5 on the reprocessing and waste management portions of the LWR fuel cycle. It contains comments by the NRC task force and the public to the notice of change, to the analysis of NUREG-0116, and to the original survey, WASH-1248. Pertinent technical data is provided in appendixes.

KEY WORDS: environment, LWR

D.11   NRC Task Force Report on Review of the Federal/State Program for Regulation of Commercial Low-Level Radioactive Waste Burial Grounds

NRC

NUREG-0217 and Suppl. 1, March 1977

This is the report of an NRC task force which examined the programs of NRC and Agreement State governments to regulate the disposal of commercial low-level waste. The issue discussed is federal vs. state regulation of commercial radioactive waste burial grounds, and related topics such as R&D, standards, and criteria. Supplement contains an analysis of the report and public comments.

KEY WORDS: low-level waste, waste management programs

D.12   The United States Program for the Safety Assessment of Geologic Disposal of Commercial Radioactive Wastes

H. Claiborne, OWI

CONF-770565; Proceedings of a Workshop on Risk Analysis and Geologic Modelling, Ispra, Italy, May 1977

This paper describes a comprehensive safety assessment program established as part of the NWTS program. The plan for achieving the objective of assessing the safety associated with the long-term disposal of high-level radioactive waste in a geologic medium is outlined.

KEY WORDS: waste management program, risk analysis

D.13   A Bioethical Perspective on Accep-
*   table Risk Criteria for Nuclear Waste Management

M. Maxey

UCRL-52320, July 15, 1977

This report focuses on the moral and ethical judgments involved in the human and environmental risks associated with nuclear waste management. There is discussion of the balance of nature, the value of technology in human life, and our responsibilities to future generations.

KEY WORDS: environment, risk analysis, bioethics

D.14   An Analysis of the Back End of the Nuclear Fuel Cycle with Emphasis on High-Level Waste Management

Jet Propulsion Laboratory, Pasadena, CA

JPL 77-59, 1977

This report examines 1) the range of policy options open for the disposition of spent fuel rods, 2) the impact of each option on the Federal high-level waste management

program, 3) the management of already existing military waste, and 4) high-level nuclear
waste management with emphasis on interagency
coordination, Federal, State, and private
sector decisions, and program consistency and
logic.

KEY WORDS: military waste, waste manage-
               ment programs

D.15   Organization for Economic Cooperation
       and Development Countries Pursue
       Geological Disposal

       P. Gera and J. Oliver, OECD

       Nuclear Engineering International
       23, (226), January 1978, p. 35

This article stresses the need for an infor-
mational demonstration program to allay
political and public anxieties. A review
of the waste management programs of several
countries is given.

KEY WORDS: waste management programs

D.16   An Analysis of Capital and Operating
       Costs Associated with High-Level Waste
       Solidification Processes

       B. Kniazewycz, TERA and
       R. Heckman, LLNL

       UCRL-80064, March 1978

An analysis has been performed to evaluate
the sensitivity of annual operating costs
and capital costs of waste solidification
processes to various parameters defined
by the requirements of proposed FWR. The
process methods and waste forms examined
were: salt cake, spray calcine, fluidized
bed calcine, borosilicate glass, and super-
calcine multibarrier.

D.17   Determination of Radioactive Waste
       Classification System

       J. Cohen and W. King, LLNL

       UCRL-52535, March 1978

Several classification systems for radioactive wastes are reviewed and a system
is developed that provides guidance on
disposition of the waste. The system has
three classes: high-level waste (HLW),
which require complete isolation from the
biosphere for extended time periods; low-
level waste (LLW), which requires contain
ment for shorter periods; and innocuous
waste (essentially nonradioactive waste),
which may be disposed of by conventional
means. A cost-benefit analysis in accord
ance with as low as reasonably achievable
(ALARA) and National Environmental Protection Act (NEPA) guidance is also presented. The environmental effects considered were limited to those involving
human exposure to radioactivity.

KEY WORDS: classification, dose

D.18   Nuclear Waste: Increasing Scale
       and Sociopolitical Impacts

       T. LaPorte, UCB

       Science 201, July 7, 1978, p. 22

This article discusses operational aspect
of radioactive waste management, outlining
the types of information necessary to estimate costs and consequences of waste
disposal programs. An index of social ex
posure to radioactive hazard is proposed
to improve the basis for policy decisions
in this area.

KEY WORDS: waste management programs

D.19   An Application of a Method for
       Comparing One-Dimensional and
       Two-Dimensional Models of a
       GroundWater Flow System

       T. Naymik, LLNL

       UCRL-52541, August 30, 1978

To evaluate the inability of one-dimensional ground-water model to interact
continuously with surrounding hydraulic
head gradients, simulations using one-
dimensional and two-dimensional groundwater flow models were compared. This

approach used two types of models: 1) flow-conserving one- and two-dimensional models, and 2) one-dimensional and two-dimensional models designed to yield two-dimensional solutions. The hydraulic conductivities of controlling features were varied and model comparison was based on the travel times of marker particles. The solutions within each of the two model types compare reasonably well, but a three-dimensional solution is required to quantify the comparison.

D.20   Risk Assessment Review Group
  *   Report to the U. S. Nuclear
       Regulatory Commission

       H. Lewis, R. Budnitz, W. Rowe,
       H. Kouts, F. Von Hippel, W.
       Loewenstein, and F. Zachariasen

This is the report of the Review Group formed to provide advice and information to the NRC on the final report of the reactor safety study, WASH-1400, "The Rasmussen Report". The Review Group noted both positive and negative aspects of the report. Among the positive notes: the general methodology and contribution to assessing the risks of nuclear inscrutable executive summary and some calculations in the body of the report.

KEY WORDS: risk analysis

D.21   Western New York Nuclear Service
       Center Study, Final Report for
       Public Comment

       DOE

       TID-28905-1, November 1978

This report presents a description of the Western New York Nuclear Service Center, responsibility delegation, potential impacts, and estimated costs of options for the future of the center. Technical options are analyzed in accordance with existing facilities. Institutional and financial responsibilities are outlined.

D.22   Social Value Considerations in
       Nuclear Waste Management

       J. Lathrop, LLNL

       UCRL-80754, 1978

The social value problems of nuclear waste management are outlined and analyzed. A decision analytic approach is presented as a methodology for determining bounds on acceptable social risk. Particular problems with this approach are discussed and ideas for solutions presented.

KEY WORDS: risk analysis

D.23   National Waste Terminal Storage
       Repositories 1 and 2, Cost Esti-
       mate Reconciliation Study, 2
       Volumes

       J. Mattern, Stearns-Rogers En-
       gineering Co., Denver, CO, and
       J. Ritchie, Kaiser Engineers,
       Oakland, CA

       ONWI-76, October 1979

The objectives of this two-phase study are to develop 1) mechanisms for characterization of cost estimates prepared for radioactive waste repositories in salt formations and 2) and understanding of the sensitivity of costs to assumptions of the various estimation processes. Phase I (Vol. I) addresses spent fuel in domed and bedded salt formations; Phase II (Vol. II) addresses high-level waste in domed and bedded salt formations and parametric evaluations for both spent fuel and high-level waste in both media.

D.24   The Economics of Mined Geologic
       Repositories

       J. Forster, TRW Energy Systems
       Planning Division, McLean, VA

       ONWI-93, December 1979

This report presents estimates of the effects on total cost and unit cost of several major alternative geologic nuclear waste repositories. Cost uncertainties are addressed qualitatively and quantitatively. A parameter analysis of key variables, including economic factors, waste form factors, host-rock factors, repository capacity, and site-specific factors, is presented.

D.25   Accident Risk Assessment - Status Report on the EPRI Fuel Cycle

Science Applications, Inc., Palo Alto, CA

EPRI NP-1128, 1979

This report was prepared for the Electric Power Research Institute (EPRI) to present the status of the EPRI project for assessing the risks of processes that support nuclear power plants. It summarizes work on the accidental radiological risk of reprocessing spent fuel, mixed oxide fuel fabrication, the transportation of materials within the fuel cycle, and the disposal of nuclear wastes It includes also work on the routine atmospheric radiological risk of mining and milling uranium-bearing ore.

KEY WORDS: risk analysis, environment, radiological consequences

D.26   State of Waste Disposal Technology and The Social and Political Implications

R. Post, ed., University of Arizona

Proceedings of the Symposium on Waste Management, sponsored by the University of Arizona College of Engineers and the Arizona Energy Commission, Tucson, Arizona, 1979

Papers included in the proceedings cover a range of topics dealing with the technical aspects of nuclear waste disposal, the treatment of wastes, criteria for evaluating waste management systems, and the special problems of waste classification applied to specific disposal sites and solid wastes. The proceedings include also reports on low-level wastes, transportation, decommissioning, and regulatory issues.

KEY WORDS: waste management programs, low-level waste, decommissioning, regulations

D.27   Workshop Proceedings on Consultation and Concurrence, Orcas Meeting, Eastsound, WA, September 1979

ONWI

ONWI-87, January 1980

The purpose of this ONWI-sponsored workshop was to probe the concepts of consultation and concurrence and to address issues in establishing and maintaining an effective state-federal process in nuclear waste management. This report presents a summary of discussions, a background paper, and papers presented on topics including state/federal perspectives, legal constraints, and strategies for nuclear waste management. Also included are remarks of several key participants.

KEY WORDS: waste management programs

D.28   Repository Sealing Design Approach - 1979

D'Appolonia Consulting Engineers, Pittsburgh, PA

ONWI-55, March 1980

This report describes a preliminary comprehensive logical design approach for sealing boreholes, shafts and tunnels to and near a deep geologic nuclear waste repository. Four alternative designs are presented, as well as considerations of seal environment, seal material, and seal geometry.

D.29 The Discount Rate in the Spent Fuel Storage and Disposal Fee

J. Forster and S. Cohen, TRW Energy Systems Planning Division, McLean, VA

ONWI-189, April 1980

This report describes a study to evaluate the suitability of alternative discount rates for use in calculating the fee the U. S. Government will charge utilities for acceptance and disposition of commercial spent nuclear fuel. The study discusses existing government guidelines and current DOE policy. Discount rates based on several different financial assumptions are evaluated and analysis results summarized.

D.30 Projected Costs for Deep Geologic Repositories for Spent Fuel Disposal

ONWI

ONWI-191, October 1980

This report presents cost estimates for the research, development, engineering, licensing, and management of design and construction of nuclear waste disposal facilities. It also contains estimated costs for design, construction, operation, and decommissioning of such facilities.

KEY WORDS: waste management programs

D.31 Decommissioning Handbook

W. Manion and T. LaGuardia
Nuclear Energy Services, Inc.
Danbury, CT

DOE/EV 10128-1, November 1980

This handbook describes all stages of the decommissioning process, including selection of the end product, estimation of the radioactive inventory, estimation of occupational exposures, the state of the art of decontamination, remote cutting of heavy metal components and structures, segmenting thick reinforced concrete structures, disposition of wastes, and estimation of program costs. Detailed summaries are given of available technologies and special construction and calculational techniques.

KEY WORD: decommissioning

D.32 NWTS Program Criteria for Mined Geologoic Disposal of Nuclear Waste - Program Objective, Functional Requirements, and System Performance Criteria

DOE

DOE/NWTS-33(1), Public Draft, April 1981

The qualitative statements in this document address the broad issues of public and occupational health and safety, institutional acceptability, engineering feasibility, and economic factors. The objectives, requirements, and performance criteria presented have been developed on the basis of DOE's analysis of what is needed to achieve safe waste disposal in an environmentally acceptable and economically feasible manner. Public review of the document is solicited.

KEY WORDS: environment, waste management programs

D.33 Analysis of Protection Requirements for Away - From - Reactor Spent Fuel Storage

A. Winbled and F. Dean, SLA

Institute of Nuclear Materials Management Annual Meeting, San Francisco, CA, July 1981

This paper describes elements of physical protection systems necessary to counter postulated threats. The report includes an analysis of the consequences of causing radiological disposal of spent fuel.

KEY WORDS: safeguards, radiological consequences

D.34   Safeguards for High-Level Waste
       Repositories – Development of a
       Technical Basis for the NRC

       J. Stoddard and L. Harris, Science
       Applications, Inc.

       Institute of Nuclear Materials
       Management Annual Meeting, San
       Franciso, CA, July 1981

This paper describes the methodology used
and the results obtained from research to
establish a technical basis for the reso-
lution of safeguards issues for high-level
waste repositories. Several alternative
high-level waste security systems were
designed and costs estimated, each corres-
ponding to different levels of protection.

KEY WORDS: safeguards, radiological
              consequences

D.35   High-Level Radioactive Waste
       Management Alternatives

       AEC

       WASH-1297, May 1974

       (SEE ABSTRACT C.12.)

D.36   Environmental Statement: Manage-
       ment of Commercial High-Level
       and Transuranium-Contaminated
       Radioactive Waste

       AEC

       WASH-1539, Draft, September 1974

       (SEE ABSTRACT C.14.)

D.37   Preliminary Assessment of the Radio-
       logical Protection Aspects of Dis-
       posal of High-Level Waste in Geologic
       Formations

       M. Hill and P. Grimwood, NRPB

       NRPB-R69, January 1978

       (SEE ABSTRACT C.63.)

D.38   Status of Nuclear Fuel Reprocessing,
       Spent Fuel Storage, and High-Level
       Waste Disposal

       E. Varanini, III, and R. Maullin,
       Nuclear Fuel Cycle Committee of the
       California Energy and Resources Con-
       servation and Development Commission
       Draft, January 1978

       (SEE ABSTRACT C.64.)

D.39   Environmental Impact Statement:
       Waste Isolation Pilot Plant,
       2 Volumes

       DOE

       DOE/EIS-0026-D, April 1979

       (SEE ABSTRACT C.125.)

D.40   Environmental Impact Statement: Man-
       agement of Commercially Generated
       Radioactive Waste, 2 Volumes

       DOE

       DOE/EIS-0046-D, April 1979

       (SEE ABSTRACT C.126.)

D.41   Regional Environmental Characteriza-
       tion Report for the Paradox Bedded
       Salt Region and Surrounding Territory

       Bechtel National, Inc.,
       San Francisco, CA

       ONWI-68, November 1980

       (SEE ABSTRACT C.181.)

E.   BIBLIOGRAPHIES

E.1   Abstracts - NRC Waste Management Program Reports, UCID-18133, 1979.

E.2   Annotated Bibliography: Hazard Assessments for the Geologic Isolation of Nuclear Wastes, B. Suta, S. Mara, S. Radding, and L. Weisbecker, November 1977, Y/OWI/SUB-77/42508.

E.3   Bibliography of Geologic Studies, Columbia Plateau (Columbia River Basalt) and Adjacent Areas in Idaho, W. Strowd, RHO-BWI-C-44.

E.4   A Bibliography on Ocean Waste Disposal, H. Stanley and D. Kaplanek, Interstate Electronics Corporation, Anaheim, CA, IEC-4460C0417, September 1976.

E.5   Decommissioning of Nuclear Facilities - An Annotated Bibliography, G. Konzek and C. Sample, BPNL, NUREG/CR-0131, October 1978.

E.6   Geological and Geochemical Aspects of Uranium Deposits, a Selected Annotated Bibliography, M. White and P. Garland, June 1978, ORNL/EIS-121/VI.

E.7   Guide to Radioactive Waste Management Literature, B. Houser, C. Holoway, and D. Madewell, ORNL, ORNL-5226, October 1977.

E.8   Information Pertinent to the Migration of Radionuclides in Groundwater at the Nevada Test Site - Part 2: Annotated Bibliography, UCRL-52078, August 1976.

E.9   Ocean Waste Disposal, A Bibliography with Abstracts, R. Brown, ed., NTIS, Springfield, VA, NTIS/PS-77/0661, and NTIS/PS-78/0706.

E.10   ONWI Library Reports Lists, published periodically by ONWI.

E.11   Radioactive Waste Disposal in Salt Deposits - A Bibliography with Abstracts, NTIS, Springfield, VA, NTIS/PS-78/1234, 1978.

E.12   Radioactive Waste Management, Abstracts published twice a month by DOE through TIC.

E.13   Radioactive Waste Processing and Disposal - A Bibliography, DOE/TIC-3311 series and -3555 series, issued periodically.

E.14   "Radioactive Wastes", C. Straub, University of Minnesota, Minneapolis, MN, Journal of Water Pollution Control Federation 47(6), 1975, p. 1498.

E.15   Resolving Community Conflict in the Nuclear Power Issue: A Report and Bibliography, R. Burt, M. Fischer, T. Corbett, K. Garrett and M. Lundgren, February 1978, Y/OWI/SUB-78/22336.

E.16   A Selected, Annotated Bibliography of Studies Relevant to the Isolation of Nuclear Wastes, ORNL/EIS-156, September 1980.

F.   JOURNALS AND PROGRESS REPORTS

JOURNALS

F.1   Radioactive Waste Management - An International Journal, D. Anderson Editor-in-Chief, SLA, Harwood Academic Publishers, New York, N. Y.

This is the first journal devoted exclusively to problems of radioactive waste management. Reports are published in the areas of 1) waste arising, inventories, collections and categorization, 2) waste treatment and conditioning, 3) transportation and interim storage, 4) terminal disposal, including site selection repository design, and risk analysis, 5) alternative management strategies and cost/benefit analysis, and 6) societal, philosophical, and political issues.

F.2   European Applied Research Reports - A Journal of European Science and Technology, Harwood Academic Publishers, New York, N. Y.

The Nuclear Science and Technology Section of this journal publishes refereed papers on all applied research areas in nuclear science and technology, including nuclear waste management, which have been sponsored by or published in collaboration with the Commission of European communities.

PROGRESS REPORTS

F.3   Basalt Waste Isolation Project Annual Report, issued annually by Rockwell International, Richland, WA

F.4   Division of Waste Management Programs Progress Report, published for DOE by Hanford Engineering Development Laboratory, Richland, WA

F.5   National Waste Terminal Waste Storage Project Monthly Technical Status Report and Annual Report, published periodically by OWI, Washington, D. C.

F.6   NTS Terminal Waste Storage Project Monthly Technical Status Report and Annual Report, published by DOE Nevada Operations Office, Las Vegas, NV

F.7   Nuclear Waste Isolation Activities Report, published occasionally by OWI,
       Washington, D. C.

F.8   Nuclear Waste Management and Transportation Progress Report, published
       quarterly by BPNL

F.9   Office of Waste Isolation Progress Report, published monthly by OWI,
       Washington, D. C.

F.10   Research and Development Semiannual Progress Report, published semiannually by
       by Rockwell International, Richland, WA

F.11   Technical Progress Report, published quarterly by ONWI, Columbus, OH

F.12   Waste Isolation Projects Progress Report, published annually by LLNL,
       Livermore, CA

GLOSSARY

Actinides - Elements of atomic number 89 to 103.

Activity - The process of making a material radioactive by bombardment with neutrons, protons, or other nuclear particles.

Adiabatic - Pertaining to the relationship of pressure and volume when a gas or fluid is compressed or expanded without heat exchange.

ALARA - As low as reasonably achievable. In the nuclear industry all radiation doses are maintained as far below prescribed standards as reasonably achievable.

Alluvial - Pertaining to or composed of alluvium, or deposited by a stream or running water.

Aquifer - A body of rock that contains sufficient saturated permeable material to conduct groundwater and to yield economically significant quantities of groundwater to springs and wells.

Aquitard - A confining bed that retards but does not prevent the flow of water to or from an adjacent aquifer.

Artesian - An adjective referring to groundwater confined under hydrostatic pressure.

Artesian head - The hydrostatic head of an artesian aquifer or of the water in the aquifer.

Attitude - The position of a structural surface relative to the horizontal, expressed quantitatively by strike and dip measurements.

Backfill - Sand, crushed rock, or other material used to fill voids in underground openings.

Bedded - Referring to rocks or salt; property resulting from consolidated sediments, with planes of separation between depositional units.

Bed load - The part of the total stream load that is moved along the stream bed.

Biosphere - All the area occupied or favorable to occupation by living organisms, including part of the lithosphere, hydrosphere, and atmosphere.

Bittern - The liquid remaining after seawater has been concentrated by evaporation until the sodium chloride has crystallized out.

Boiling water reactor (BWR) - A reactor system that uses a boiling water primary cooling system. Primary cooling system steam turns the turbines to generate electricity.

Borehole - A hole drilled into soil or rock to obtain geologic and hydrologic information.

Breccia - A coarse-grained clastic rock composed of large, angular, and broken rock fragments that are cemented together in a finer-grained matrix.

Breccia pipe - A vertical column of brecciated rock, formed by the upward-moving dissolution of rock material.

Buffer - Material placed between a canister, overpack, or sleeve, and the geomedium.

Canister - A container for spent fuel and/or solid high-level waste.

Cementation - The diagnetic process by which clastic sediments become lithified or consolidated into hard, compact rocks through the disposition of precipitation of minerals in the space between the individual grains of the sediment.

Clastic rock - A consolidated sedimentary rock composed of broken fragments, which are derived from preexisting rocks and which have been transported individually for some distance from their place of origin.

Clay seam - A thin layer of clay separating two distinctive layers of different composition or of greater extent.

Curie (Ci) - A unit of measure of radioactivity; 1 curie = that quantity of any nuclide which undergoes 3.7×10^{10} disintegrations per second.

Craton - A part of the earth's crust, which has attained stability, and which has been little deformed for a prolonged period.

Creep - A slow deformation in rock or salt that results from the application of constant stress.

Decommissioning: Preparation of worn-out or obsolete nuclear facilities for retirement. Decommissioning operations remove facilities such as reprocessing plants and burial grounds from service and reduce or stablize radioactive contamination. Concepts include:

      demolition and restoration to original conditions requiring no control,

      partial demolition and fixation of residues,

      minimal demolition followed by isolaton and control of residues.

Denudation - The sum of the processes that result in the wearing away or the progressive lowering of the earth's surface by various natural phenomena, including weathering, erosion, mass-wasting, and transportation.

Deposition - The placing or throwing down of any material, specifically the constructive process of accumulation into beds, veins, or irregular masses of any kind of loose, solid, rock material by any natural agent, such as the mechanical settling of sediment from suspension in water, the chemical precipitation of mineral matter by evaporation from solution, or the accumulation of organic material through the processes of death of plants and animals.

Depository — A subterranean cavern excavated in a deep geologic medium for the disposal of nuclear waste.

Detritus — A collective term for loose rock and mineral material that is worn off or removed directly by mechanical means such as disintegration or abrasion.

Diagenesis — All the chemical, physical, and biological changes or transformations, exclusive of weathering and metamorphism, undergone by a sediment after its initial deposition and during and after its lithification.

Diapirism — The process of piercing or rupturing of domed or uplifted overlying rocks by core material in a plastic state.

Dilation theory — The theory that attributes glacier movement to infiltration and freezing of water in cracks and other openings.

Dip — The angle at which a planar feature is inclined to the horizontal.

dissolution — A space or cavity in or between rocks, formed by the breakdown or dissolving of part of the rock material.

Ecology — The study of relationships between organisms and their environments, including the study of communities, patterns of life, natural cycles, relationships of organisms to each other, biogeography, and population changes.

Ecosystem — A unit in ecology consisting of the environment with its living elements and the factors which exist in and affect it.

Effective porosity — The portion of the pore space in a saturated permeable material in which flow of water takes place.

Eh — Oxidation potential of a system with respect to the normal hydrogen electrode of a standard hydrogen half-cell.

Evapotranspiration — Loss of water from a land area through transpiration of plants and evaporation from the soil.

Facies — The sum of all primary lithologic characteristics exhibited by a sedimentary rock and from which its origin and environment of deposition may be inferred.

Fault — A surface or zone of rock fracture along which there has been displacement.

Fissile material — One of several actinides which under proper conditions fission spontaneously, producing sufficient neutrons to sustain a chain reaction.

Fission (nuclear) — The splitting of a heavy nucleus into two or, rarely, more fragments.

Fissionable material — Any material fissionable by neutrons of all energies, such as certain isotopes of uranium and plutonium.

Flood plain — The surface of relatively smooth land adjacent to a river channel, constructed by the present river in its existing regimen and covered with water when the river overflows its banks at times of high water.

Fluvial erosion — Erosion of the land surface by the scouring action of streams and rivers and the action of rainwater flowing over the ground surface.

Fuel (nuclear reactor) - Fissionable material used as the source of power when placed in a critical arrangement in a nuclear reactor.

Fuel cycle - The complete series of steps involved in supplying fuel for nuclear reactors. It includes mining, refining, enrichment, fabrication of fuel elements, use in a reactor, chemical processing to recover the fissionable material remaining in the spent fuel, enrichment of the fuel material, refabrication of new fuel elements, and management of radioactive waste.

Fuel element - A tube, rod, or other form into which fissionable material is fabricated for use in a reactor.

Fuel reprocessing plant (FRP) - Plant where irradiated fuel elements are dissolved, waste materials removed, and reusable materials are segregated for reuse.

Galvanic attack - Erosion or corrosion of materials due to electric currents.

GEIS - Generic Environmental Impact Statement

Geochemistry - The study of the distribution and amounts of the chemical elements in minerals, ores, rocks, soils, water, and the atmosphere, and the study of the circulation of the elements in nature, on the basis of the properties of their atoms and ions.

Geology - The science that deals with the history of the earth and its life, especially as recorded in rocks.

Geomorphology - The study of the classification, description, nature, origin, and development of present landforms and their relationships to underlying structures and geologic history.

Glacial drift - A generic term for rock material transported by glaciers and deposited on land or in the ocean.

Groundwater - That part of the subsurface water that is the zone of saturation; used loosely, all subsurface water, excluding internal water.

Halite - Rock salt.

Head - See hydrostatic head.

Heat flow - The product of thermal conductivity of a substance and the thermal gradient in the direction of flow of heat.

High-level liquid waste (HLLW) - The aqueous waste resulting from operation of the first cycle solvent extraction system (or its equivalent) in a facility for reprocessing irradiated reactor fuels as well as concentrated wastes from subsequent cycles.

High-level waste (HLW) - DOE management directives define high-level waste to include high-level liquid wastes, products from solidification of high-level liquid waste, and irradiated fuel elements if discarded without reprocessing. A proposed NRC regulation (10 CFR 60.3) defines high-level waste to include irradiated fuel, high-level liquid waste, and products from its solidification. In the GEIS there are instances, however, where discarded spent fuel and high-level waste (as wastes from the reprocessing of spent fuel) are cited separately.

Horizon - A plane of stratification (e.g., a stratum) assumed to have once been horizontal and continuous.

Hydraulic gradient - In an aquifer, the rate of change of pressure head per unit of
distance of flow at a given point and in a given direction.

Hydrology - The science that deals with continental water, its properties, circulation,
and distribution, on and under the earth's surface and in the atmosphere.

Hydrostatic head - The height of a vertical column of water, the weight of which for
a unit cross section is numerically equal to the hydrostatic pressure at a point.

Hydrostatic level - The level to which the water will rise in a well under its full
pressure head.

Hydrostatics - That aspect of hydromechanics which deals with forces that result in
equilibrium.

Hypsithermal - A term proposed as a substitute for climatic optimum and thermal
maximum, and representing the postglacial interval when most of the world
entered a period when mean annual temperatures exceeded those of the present.

Incision - The process whereby a downward-eroding stream deepends its channel or
produces a narrow, steep-walled valley.

Interstitial - Pertaining to a mineral deposit in which the minerals fill the pores of
the host rock.

Isopleth - A line or surface on which some mathematical function has a constant value.

Isothermal - Pertaining to the process of changing the thermodynamic state of a substance, e.g., its pressure and volume, while maintaining constant temperature.

Joint - A surface of actual or potential fracture or parting in a rock, without
displacement.

Leaching - The separation, selective removal, or dissolving out of soluble constituents
from a rock or ore body or material by the action of percolating water.

Ligands - The atoms, radicals, or molecules that are bound to the characteristic or
central atom of a polyatomic group in inorganic compounds.

Light water reactor (LWR) - May be either a BWR or PWR; uses as coolant ordinary water
(H_2O) instead of heavy water (D_2O).

Liquefaction - The transformation of loosely packed sediment into a fluid mass.

Lithic - Pertaining to or made of stone.

Lithology - The physical characteristics of a rock.

Lithosphere - The solid portion of the earth, as compared with the atmosphere and the
hydrosphere.

Magma - Naturally occurring mobile rock material, generated within the earth and capable
of intrusion and extrusion, from which igneous rocks are thought to have been derived
through solidification and related processes.

Mass wasting - Generic term for the dislodgement and downslope transport of soil and
    rock material under the direct application of gravitational body stresses.

Matrix - The smaller or finer-grained, continuous material enclosing or filling the inter-
    stices between the larger grains or particles of a sediment or sedimentary rock;
    the natural material in which a sedimentary particle is embedded.

Mineralogy - The study of minerals, including formation, occurrence, properties, compo-
    sition, and classification.

Outcrop - That part of a geologic formation or structure that appears at the surface of
    the earth.

Overpack - An integral closed container or containers in which a canister is placed.

Package - All of the material emplaced in a hole drilled in the geomedium. It may include
    a canister, overpack, sleeve, buffer, and air gap.

Paleontology - The study of life in past geologic periods based on fossil plants and
    animals and including phylogeny, relationships to existing plants and animals,
    and the chronology of the earth.

Permeability - The property or capacity of a porous rock, sediment, or soil for trans-
    mitting a fluid without impairment of the structure of the medium.

Permeability coefficient - The rate of flow of water in gallons per day through a cross
    section of one square foot under a unit hydraulic gradient, at the prevailing temp-
    erature or adjusted for a temperature of 60°F.

Petrology - That branch of geology dealing with the origin, occurrence, structure, and
    history of rocks, especially igneous and metamphoric rocks.

pH - Negative logarithm of the hydrogen ion activity.

Piezometric head - See hydrostatic head.

Porosity - The ratio of the volume of voids in a rock or soil to its total volume.

Post-closure - The period after the decommissioning of a nuclear repository.

Pressurized water reactor (PWR) - A reactor system that uses a pressurized water
    primary cooling system. Steam formed in a secondary cooling system is used
    to turn turbines to generate electricity.

Rem - (roentgen equivalent man) - A quantity used in radiation protection to express
    the effective dose equivalent for all forms of ionizing radiation. It is the
    product of the absorbed dose in rads and factors related to relative biologi-
    cal effectiveness.

Repository - The entire region, subterranean and surface, which will be owned by the
    Federal Government for the purpose of high-level waste disposal.

Retrieval - Removal of an emplaced canister and its overpack from the sleeve and its
    return to the repository surface facility.

Rock salt - Coarsely crystalline halite occurring as a massive, fibrous, or granular aggregate, and constituting a nearly pure sedimentary rock that may occur in domes, plugs, or as extensive beds resulting from the evaporation of seawater.

Roentgen - A unit for measuring gamma or "x-ray" radiation. The roentgen is defined by measuring the effect of the radiation on air. It is that amount of gamma or x-rays required to produce ions carrying 1 electrostatic unit of charge in 1 cubic centimeter of dry air under standard conditions.

Sediment - Solid fragmented material, or a mass of such material, either inorganic or organic, which originates from weathering of rocks and is transported by, suspended in, or deposited by air, water, or ice, or precipitation from solution or secretion by organisms, and which forms in layers on the earth's surface at ordinary temperatures in a loose, unconsolidated form, e.g., sand, gravel, slit, mud, till, alluvium.

Seismology - The study of earthquakes; by extension, the study of the structure of the interior of the earth via both naturally and artificially generated signals.

Sleeve - A can, usually open at the top, into which the canister and its overpack are placed. The top may be closed by a plug. Its purpose is to protect the contents from geostresses and permit ready retrievability.

Solidification - Conversion of liquid radioactive waste to a dry, stable solid.

Spent fuel (SF) - Nuclear reactor fuel that has been used to the extent that it can no longer be used in a nuclear power plant.

Soret coefficient - A dimensionless thermal diffusion coefficient involved in coupled heat and mass transfer analysis.
Soret effect - Diffusion or transport due to temperature difference exclusively.

Strain - Linear or volumetric deformation resulting from an applied stress, expressed as a percentage or fraction of the original length or volume.

Stratigraphy - The branch of geology that deals with the definition and description of major and minor natural divisions of rock (mainly sedimentary strata available for study in outcrop or from subsurface).

Stratum - A tabular or sheet-like mass (layer) of homogeneous or gradational sedimentary material or any thickness, visually separable from other layers above and below; a a sedimentary bed.

Stream bed - The channel containing or formerly containing the water of a stream.

Stream load - The solid material that is actually transported by a stream, either as visible sediment or in chemical or colloidal solution.

Stress - In a solid, the force per unit area, acting on any surface within it.

Strike - The direction or trend that a structural surface, e.g., a bedding or fault plane, takes as it intersects the horizontal.

Tectonics - A branch of geology dealing with the broad structure and deformational features of the upper part of the earth's crust.

Thermal conductivity, coefficient of - The time rate of transfer of heat by conduction, through a unit thickness, across a unit area, for a difference of temperature of one degree centigrade.

Thermal diffusivity, coefficient of - Coefficient of thermal conductivity divided by the product of the density and the specific heat capacity of the material.

Transuranic (TRU) elements: Elements with mass number greater than 92. They include, among others, neptunium, plutonium, americium, and curium.

Transuranic waste - Waste material measured or assumed to contain more than a specified concentration of transuranic elements. For purposes of this statement, TRU waste is currently identified as containing 10 nanocuries or more of transuranic activity (usually ^{239}Pu) per gram of waste.

Uraninite - A strongly radioactive octahedral or cubic mineral, essentially UO_2, but usually partly oxidize; the chief ore of uranium.

Venting - The escape through the earth to the atmosphere of gases or radioactive products from an underground high-explosive or nuclear detonation. In general, escape of gas from any container.

Volcanism - The processes by which magma and its associated gases rise into the crust and are extended onto the earth's surface and into the atmosphere.

Volcanology - The branch of geology that deals with volcanism, its causes and phenomena.

Waste form - The nuclear waste residue configuration, which is sealed within a canister.

Waste package - The waste form, canister, overpacks, absorbent materials, or other engineered barriers placed around the canister.

Weathering - The destructive process or group of processes whereby earthy and rocky material, on exposure to atmosphere agents at or near the earth's surface, are changed in character, with little or no transport of the altered material; part of erosion.

Zeolite - A generic term for a large group of white or colorless, sometimes red or yellow, hydrous aluminosilicates.

AUTHOR INDEX

Acres American, Inc., C.177
Alexander, C., A.12
Allen, E., C.86
Altenbach, T., C.128
American Nuclear Society, C.18
Ames, L., C.78
Amick, C., C.132
Anderson, D., C.43
Anderson, T., C.98
Andersson, B., C.83
Andersson, L., C.131
Angerman, C., A.7
Angino, E., C.9, C.36, C.37, C.54
Apps, J., C.75
Arbital, J., C.146, C.174
Atcher, R., C.41
Atlantic Richfield Hanford Company, C. 13
Atomic Energy Commission, C.12, C.14, D.2,
    D.4

Baker, D., A.3
Barker, J., C.127
Ballou, L., C.133
Basin, S., C.182
Battelle Pacific Northwest Labs, C.194
Bechtel National, Inc., A.31, C.179,
    C.180, C.181
Bell, M., C.5
Berbano, M., C.74
Bertozzi, G., C.16, C.99, C.198
Bettis, E., C.146
Bibler, N., A.21
Binall, E., C.108
Blomeke, J., A.12
Bondietti, E., C.122
Bopp, C., C.51
Boulton, J., C.89
Bowers, V., C.42
Bradley, D., C.53
Bredehoeft, J., C.102
Brookins, D., C.97
Brough, W., C.133
Brown, R., E.9
Brown, W., C.177
Brunton, G., C.52

Buckner, J., B.8, C.183
Buckner, M., B.8, C.183
Budnitz, R., D.20
Burkholder, H., A.3, C.144
Burleigh, R., C.108
Burt, R., E.15
Butkovich, T., C.133
Butler, J., C.177

California State University, Hayward, CA,
    C.160
Campbell, J., C.91, C.92
Canadian Department of Energy, Mines and
    Resources, C.67
Carlson, R., C.133
Carlsson, H., C.81, C.84, C.140
Carpenter, D., C.124
Carruthers, J., C.24
Casey, L., A.24
Chan, T., C.87, C.193
Cheung, H., A.28
Claiborne, H., C.8, D.12
Clarke, R., B.1
Cloninger, M., A.3, C.173
Cohen, B., C.38
Cohen, J., A.20, D.9, D.17
Cohen, S., D.29
Commission of European Communities, C.40
Committee on Radioactive Waste Management,
    National Research Council, A.34,
    C.85, C.106
Comper, W., C.25
Cook, N., C.75, C.79, C.87, C.90. C.96,
    C.163, C.193
Corbett, T., E.15
Cottrell, D., C.145
Cowan, G., C.55
Crandall, D., C.152
Crandall, J., C.134
Croff, A., A.12

D'Allessandro, M., C.99, C.198
Dames and Moore, Inc., C.103, C.177
D'Appolonia Consulting Engineers, Inc.,
    C.109, C.170, D.28